THE INVENTION OF TOMORROW

Also by Thomas Suddendorf

The Gap: The Science of What Separates Us from Other Animals

THE
INVENTION OF
TOMORROW

A NATURAL
HISTORY
OF FORESIGHT

THOMAS SUDDENDORF,
JONATHAN REDSHAW, AND
ADAM BULLEY

BASIC BOOKS

New York

Basic Books
Hachette Book Group
1290 Avenue of the Americas, New York, NY 10104
www.basicbooks.com

Printed in the United States of America

First Edition: September 2022

Published by Basic Books, an imprint of Perseus Books, LLC, a subsidiary of Hachette Book Group, Inc. The Basic Books name and logo is a trademark of the Hachette Book Group.

The Hachette Speakers Bureau provides a wide range of authors for speaking events. To find out more, go to www.hachettespeakersbureau.com or call (866) 376-6591.

The publisher is not responsible for websites (or their content) that are not owned by the publisher.

Additional credits information for figures is on page 209.

Print book interior design by Amy Quinn.

Library of Congress Cataloging-in-Publication Data has been applied for.

ISBNs: 9781541675728 (hardcover), 9781541675735 (ebook)

LSC-C

Printing 1, 2022

For Michael C. Corballis
—T. S.

For My Family
—J. R.

For Dad
—A. B.

CONTENTS

CHAPTER 1

YOUR PRIVATE
TIME MACHINE

Over five thousand years ago, a man braced himself against the bitter cold as he ascended a mountain in the Alps. He was badly hurt. An attack in the valley below had left him with a deep stab wound between the thumb and index finger of his right hand. Now he was far from home, at least one or two days' walk through the ice and forest. This was not a journey for the faint of heart—or for the unprepared. The man wore a coat, leggings, hat, shoes, backpack, and other attire that had been stitched together from the skins of several different animals. In a pouch on his belt he held stone tools for cutting, scraping, and boring, as well as chunks of pyrite he could strike to make fire. Threaded onto a leather strap were two chewy lumps of antiparasitic birch fungus fruit—useful when you're suffering from whipworm. He had a copper axe, bow and arrows, and a dagger made of chert. The dagger had been painstakingly sharpened many times. He might need it. Was he being followed, even up here in the snow at the top of the mountain?

The man's frozen, mummified corpse was found in 1991 preserved in an Alpine glacier, surrounded by his extensive equipment. He came to be known as Ötzi, after the Ötztal Alps where he was found, or simply as "the

Iceman."* Ötzi's preparations illustrate the power of our species' universal capacity to remember what worked in the past and think ahead to what might be needed tomorrow.[1]

The human mind is a virtual time machine. With it we can relive past events and imagine future situations, even if we have never experienced similar situations before. Humans do this incessantly, daydreaming about summer vacations, savoring the thought of upcoming dinner dates, and brooding over test results. Because humans are mental time travelers, we can prepare for opportunities and threats well in advance, as Ötzi did, trying to shape the future to our own design. Foresight, our ability to anticipate events and act accordingly, is perhaps the most powerful tool at our disposal. This book is about the nature and evolution of this ability and its role in the human story.[2]

Of course, just because we can imagine the future does not mean we actually know what will happen. In the end, Ötzi probably didn't foresee the arrow that hit him in the back and put him on ice for five thousand years. Much of what comes to pass we do not anticipate, and much of what we anticipate does not come to pass. Even professionals specializing in prediction, such as stockbrokers and meteorologists, often struggle to forecast the price of gold next quarter or whether it will rain next Tuesday. Human foresight can fail spectacularly. You may have heard of backyard engineers attaching helium balloons or rockets to their chairs in eager anticipation of flight or speed but without adequate contemplation of how they might suddenly drop or stop.[†3]

---

* The journey back through the millennia to uncover details of Ötzi's preparations has been one of the great interdisciplinary detective stories of recent decades. We can know what he ate because of the food left inside his digestive tract, what route he took in his final hours from the pollen on this food, and what animals his clothes were made of from remnant mitochondrial DNA. We can infer that he was attacked because hand injuries like his are commonly the result of trying to block a blade, and that his tools were well loved from use-wear analysis of their surfaces.

† Famously, in 1982 Larry Walters attached forty-two helium weather balloons to his lawn chair to take flight. He did think ahead sufficiently enough to bring an air pistol on the journey, allowing him to pop balloons to get back down, even though he utterly miscalculated the speed and reach of his takeoff as he soared to a height of sixteen thousand feet.

There are many good reasons to complain about how humans pilot their mental time machines. For one thing, we are often lousy at taking a longer view, being guided instead by the prospects of a quick buck, the whims of the daily news cycle, or likes on social media. We persistently predict our projects will be finished within budget and on time, even if our rosy forecasts have frequently been wrong before. And we tend to expect that negative events, say, falling off a ladder, are less likely to happen to us than they actually are.[4] Human history is littered with anecdotes of poor planning with catastrophic consequences, such as when Queensland government officials brought cane toads to Australia to kill the pesky cane beetle, only for the toads to reproduce out of control and ravage the local ecosystems. So this book is also about the many limits of human foresight, and how people deal with them.

Throughout history, humans have conjured up audacious strategies to help them peek ahead in time. An entire alphabet's worth of fortune-telling methods abounds, from *abacomancy*, reading the future in dirt, sand, smoke, or ashes, to *zoomancy*, reading it from the behavior of birds, ants, goats, or asses. What these "-mancies" have in common, of course, is that they do not work as advertised. Nonetheless, they are among the many products of a fundamental human quest: to grapple with an uncertain future and figure out the best way forward.

While the future cannot be found in entrails or tea leaves, there are some patterns in nature that can help us predict and prepare. The ancient Greeks, though they would routinely consult the oracle before embarking on a major venture, also created remarkably effective forecasting tools. A room in the Greek National Archaeological Museum in Athens is dedicated to one particularly enigmatic artifact used for this purpose. Pulled from the Aegean Sea in 1901 by sponge divers off the island of Antikythera, the unassuming lump of mangled wood and corroded metal would only many decades later be identified as the world's earliest-known analog computer. It is over two thousand years old.[5]

The Antikythera mechanism is a relic of astonishing technological complexity, featuring dozens of interlocking bronze gears and faded, arcane inscriptions. By turning a hand crank, its operator could select a calendar day

Figure 1.1. The Antikythera mechanism as it looks today (*left*) and a model of what the cosmos display would have looked like (*right*).

on a front dial and predict the future of celestial bodies: the movement of the planets, phases of the moon, and eclipses of the sun. The Roman statesman Cicero enthused that by contemplating the predictable regularities of the heavens, "the mind extracts the knowledge of the Gods."[6]

Modern humans have extracted ever more knowledge about nature and how to predict its course. Though we might struggle to conduct any of the required calculations ourselves, today we can precisely forecast the time of high tide or the passage of celestial events by consulting devices that fit into our pockets. A quick look at Wikipedia tells us that Venus will transit the sun on March 27—and Mercury will do the same a day later—in the year 224,508 AD. Closer to home, our everyday lives are increasingly built on shared schedules and models of the future that guide human cooperation. We clock into our nine-to-fives, meet for weekly book clubs, and toil towards important deadlines.

Nonetheless, it is painfully obvious that even when we have a clear view of what lies ahead, we can fail to act accordingly. On Christmas Eve 2019, New York politician Brian Kolb published a newspaper column warning the public about the dangers of drunk driving—advising that by "thinking ahead and coming up with a plan before imbibing, many regrettable situations can be avoided"—only to be himself found inebriated at the wheel of his car in a ditch a week later.[7] Though it is easy to laugh at such hypocrisy, it may not be hard to come up with your own personal examples of imprudent decisions in spite of unambiguous forecasts and best intentions. When waking up with a terrible hangover, have you ever sworn never to touch a drop again—only to find yourself beer in hand before long? Have you ever ordered a greasy hamburger or an extra-large sundae despite knowing you would regret it? And then duly regretted it? Or have you ever set a New Year's resolution and discarded it weeks later, resolving to try again next year? Most of us are far from consistent in our actions, coherent in our plans, or reliably guided by rational analysis and resolve.

In this book, we tell the story of how foresight radically transformed our ancestors from unremarkable primates confined to the tropics of Africa to creatures that hold the destiny of an entire planet in their hands. But it is not simply an ode to our successes of prediction or a lament for our failures. Humans have a remarkable capacity to traverse the spans of ages in the mind's eye, but perhaps our greatest powers come from a humbler source. We understand we can't know for sure what the future holds, and realize we'd better do something about it. Paradoxically, much of the power of foresight derives from our very awareness of its limits.

———

In July 1969, shortly before Apollo 11 was about to reach the moon, President Nixon's speechwriter, William Safire, drafted the following text to be read on TV:

- -

IN EVENT OF MOON DISASTER:

Fate has ordained that the men who went to the moon to explore in peace will stay on the moon to rest in peace.

These brave men, Neil Armstrong and Edwin Aldrin, know that there is no hope for their recovery. But they also know that there is hope for mankind in their sacrifice. . . .

Others will follow, and surely find their way home. Man's search will not be denied. But these men were the first, and they will remain the foremost in our hearts.

For every human being who looks up at the moon in the nights to come will know that there is some corner of another world that is forever mankind.[8]

Of course, Neil Armstrong and Buzz Aldrin did make it home safely from the moon in 1969, and so President Nixon never had to address the nation with these words. He also never had to follow through with other harrowing aspects of the contingency plan, such as an instruction to phone the "widows-to-be" while the men remained stranded on the lunar surface. NASA has hundreds of pages of *if-then* plans, with detailed instructions for mission control on what to do in incidents ranging from a "close call" to a "type A mishap."[9]

You'll also be no stranger to contingency planning. Though we may all have a Plan A, say for our careers, we also understand that events may turn out differently from what we imagined: our company could go bust, we might get bored, or we could be hit by a bus. So we put money aside for a rainy day, keep an eye on other opportunities, and purchase comprehensive life insurance packages. People sign prenups and set up fire extinguishers for when they might be needed, all the while hoping they never will be.

In a series of studies from our research group, based at the University of Queensland in Brisbane, Australia, we gave children a simple task to assess the origins of this basic insight that the future may not turn out as predicted.[10] We dropped a reward into a vertical tube with two exits at the bottom, like an upside-down Y. In preparation for the drop, two-year-old children tended to only cover one or the other exit, which meant they only caught the reward some of the time. But by age four, most children caught on: they immediately and consistently covered their bases and held one hand under each of the exits, ensuring they would catch the prize regardless of

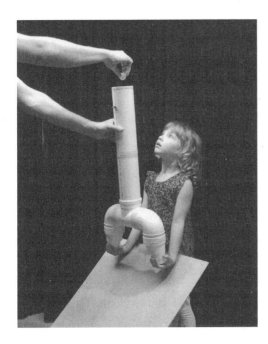

Figure 1.2. Thomas's then four-year-old daughter, Nina, hedges her bets, holding out two hands to make sure she catches the falling prize.

where it fell. Even preschoolers can compensate for the fact they cannot predict out of which exit the target will fall. They seem to know that they do not know exactly what the future will hold.

Because humans can conceive of multiple versions of the future branching from the present, we can compare our options to select the best one.[11] This capacity has far-reaching implications, as we will see throughout this book. For one thing, it gives us our intuitive sense of free will—our (some would say fanciful) impression that we are the masters of our own destiny. People tend to treasure this notion. While it's not always obvious which path is best, it is nonetheless empowering to think we are the ones behind the wheel.

The flip side of this freedom is that we know we can't wind back the clock. We are responsible for the consequences of our actions. Once we realize that our behavior could cause harm down the road, most people feel compelled to chart a different course. And, if we do end up causing harm, we often take our mental time machines for a spin into an alternative timeline, imagining what we should have done instead.[12] That's why we feel regret. In turn, we can consider what we should do next time around, and even wonder whether, one day, we might look back on our current decisions with regret and wish we'd done something else.

The sense of free will bestowed by our mental time machines not only leads us to feel accountable for our own actions but also drives us to dish out judgment, punishment, and retribution for others' actions. In many cultures, people carve up behavior into right and wrong based on what a decision-maker could have foreseen, and what options they had available at the time. If you had a choice and could have easily anticipated the negative impact of your actions on others, people tend to judge you more harshly. Lawyers, priests, and gossip columnists argue endlessly about good intentions, miscalculations, and unforeseen circumstances that increase or mitigate guilt. In common law, *mens rea* (Latin for "guilty mind") refers to the intention behind a misdeed, and establishing its existence is often essential for an action to count as a crime. Foresight is also critical when distinguishing *recklessness*, knowingly exposing others to the possibility of future harm, and *negligence*, failing to anticipate an avoidable danger that should have been foreseen.

Foresight confronts us with many conundrums, including moral dilemmas. Should we tell someone a white lie because it will comfort them? Is the suffering of lab rats worth the potential benefits of trialing a chemotherapy drug? Humans might be the only species that can run a zoo, because we can foresee what other animals need and what they are capable of, but we are also the only species that can ask if we ought to run a zoo. Humans incessantly debate what should be done to create a better future.

Unfortunately, because we are aware of tomorrow but limited in our capacity to predict what it will bring, there are endless possibilities to get worked up about—*Will lightning strike, the boat sink, or a shark attack?* The price of foresight includes worry—*How am I going to pay for it? Will they cheat on me? Is this rash just a rash?*—and even worries about worrying—*What if I can't stop and my worrying drives me mad?* Dwelling on possible future scenarios can make us anxious and leave us feeling hopeless. Plus, wish as we might that it were otherwise, some negative future events are pretty much certain. Foresight confronts us with dreadful insights, including perhaps the most unwelcome of all: that we are going to die. But even here, humans contemplate the uncertainty around what, if anything, happens afterwards. Our time-traveling minds voyage beyond our own deaths. Will we be greeted by Saint Peter at the pearly gates or by Lucifer in hell? Reincarnated as a tick or a tiger? Or will our consciousness be forever obliterated as our brain cells die? Whatever your beliefs, you are likely also to be concerned about the world you will leave behind. What will your legacy be?

Foresight is essential to the evolution and workings of the human mind, its powers and pitfalls, and to the functioning of human societies. Charles Darwin was so impressed by this ability that he was most compelled to believe in God when pondering how it could have emerged. He ruminated over the difficulties "of conceiving this immense and wonderful universe, including man with his capacity of looking far backward and far into futurity, as the result of blind chance or necessity."[13] Let's take a closer look at some of the fundamentals of this mental machine that can ferry us through time.

If a real time machine were going to be invented someday, wouldn't we be inundated by visitors from the future right now? The late great physicist Stephen Hawking once hosted a party for time travelers at Cambridge University by sending out invitations only after the day of the event. Nobody came. Perhaps ours is not a particularly interesting or pleasant time to visit. Alternatively, the absence of time tourists could be a lesson that voyages across the fourth dimension (at least into the past) will forever remain impossible. Alas, that is what physicists tell us. We can travel in time only in our minds.[14]

Imagine you are on a bus on your way home from work. As you look out the window and watch the world go by, your mind starts to wander to a job you need to complete by the end of the week. You may picture the tasks ahead and briefly recall all the trouble you had making progress today. You feel a little joy in anticipating finally getting it all off your plate. It will make the weekend so much more fun. You should celebrate. This thought reminds you of a friend's upcoming birthday and the need to organize something special. You recall the last time you organized a party for her and the fun

Figure 1.3. Professor Hawking's invitation still stands.

everyone had—except that there was not enough champagne. You make a mental note to definitely drive to the store on Saturday morning—they advertised a deal on French wines. Or has that offer already expired? Anyway, it will be great to exhale and clink glasses with friends after all that focus on work the last few weeks. As your stop comes into view, your thoughts return to the here and now; you grab your bag and get ready to hop off.

Your mind readily darts between past, present, and potential future events. It recognizes and explores the causal chains that link them as you experience hope, joy, anxiety, and many other emotional reactions to mere mental scenarios. As your eyes glide over these lines, written at an altogether different time and place, you may be carried to future parties you still have to plan, or to a recent rough day at work. How about you briefly revisit the last birthday party you attended? Your current situation, such as the sights and sounds around you, fade into the background as your mind traverses back to that experience. This is called *episodic memory*: your memory for the *episodes* of your life. Similarly, why not now ponder what you might do on your next birthday? This ability to imagine a specific future episode, and act accordingly, is what we call *episodic foresight*.[15] We'll look at each in turn.

In one of the worst cases of amnesia known to science, the English pianist, conductor, and scholar Clive Wearing suffered brain damage from a herpes infection that has left him unable to recall a single event from his past. Despite still being lucid and intelligent, Clive cannot form new memories and can track the flow of events only in the order of seconds. His wife, Deborah, has written a haunting memoir, *Forever Today*, in which she describes how Clive incessantly "wakes up," believing he is— right this moment—conscious for the first time.[16] Clive insists that his life until now has been "like death": no sights, no sounds, no thoughts, no dreams. Nothingness.

Not all memory involves mental travels to past events. Psychologists distinguish different kinds of memory systems because it is possible to lose access to your personal past while keeping hold of other information. Clive Wearing retains motor skills, such as how to play piano, a form of *procedural memory*. He also still remembers factual knowledge, such as the name of his wife, which is called *semantic memory*. Yet Clive has lost the ability to mentally travel back in time, to reexperience events that happened to him. He has lost all of his episodic memories. He knows he has children, but he cannot remember any episodes from the years he spent raising them: not their first days of school, not any of their birthday parties. Clive can play the piano, and he knows what a piano is, but he has no recollection of ever having played one in his life.

One of the most obvious functions of our virtual time machine is that it gives us a sense of continuity by enabling us to revisit the events of our lives and to draw narrative connections between them. We weave a story about ourselves from the threads of experience. Yet, curiously, a persistent finding from over a century of research is that even if we don't have a memory disorder, the accuracy of our recollections, just like many of our predictions, leaves much to be desired.[17] Our recall tends to be patchy and lacking in detail. We are often demonstrably wrong. What's more, the confidence we have in the accuracy of our memories is a surprisingly poor predictor of the actual accuracy of those memories. These are profound concerns given how much humans stake on reliable recall.

On the night of March 2, 1995, Michael Gerardi and Connie Babin exited a New Orleans restaurant after their first date. Minutes later, the pair were approaching Gerardi's car when they were accosted by three teenagers. Babin ran from the scene but briefly turned around to see one of the assailants shoot Gerardi in the head, killing him instantly. Babin would go on to identify sixteen-year-old Shareef Cousin as the killer, telling the court, "I will never forget that face." Cousin was found guilty and sentenced to execution, spending two years on death row before his conviction was ultimately overturned. In spite of Babin's confidence, Cousin was nowhere near the scene at the time of the murder and was instead being driven home by his basketball coach after a game.[18]

A long research tradition examining eyewitness testimony has shown that our recollections of events are based on active reconstruction rather than on a precise recording of past experiences.[19] We do not carry in our minds a video camera that faithfully captures our past and enables crisp replay. We actively rebuild the episodes when it is time to recall them. In the process, we may draw on information we acquired only after the event—such as what we heard others say or information that was implied in a question. When being asked whether a man's mustache was curled up or down, we may absorb the idea of a mustache into our memory of him even if there was none at all. We are prone to embellish reports in order to make a better story or to enhance or excuse our role in the unfolding of events. And, with repeated retelling, it becomes increasingly difficult to disentangle the original event from what was added afterwards.[20]

Evidently, memory is not a system for incorruptible record keeping. Judges and historians no doubt would have preferred something more reliable. From an evolutionary perspective, however, natural selection does not care if your memory is accurate looking backward. What matters are the consequences of your memory for survival and reproduction moving forward. Memory is for the future.[21]

Consider the basic form of memory known as Pavlovian conditioning. Ivan Pavlov, the Nobel Prize–winning Russian physiologist, found that his dogs learned to associate the sound of a bell with the arrival of food, such that whenever they heard that sound, they started salivating even before the food arrived. Though this is memory, of course, the process is really future directed. The dog salivates in preparation for the arrival of something to digest. Similarly, when a behavior is followed by a reward, an animal will become more likely to repeat that behavior than when it is followed by punishment. A dog learns to expect that it will be rewarded for coming when its name is called, and to expect to be punished if it jumps on the dinner table.

Humans can likewise be conditioned. Even people like Clive Wearing, who cannot recall events, can still learn in this way. The Swiss neurologist Édouard Claparède once pricked one of his amnesia patients using a tack hidden in the palm of his hand during a handshake.[22] Afterwards, despite having no recollection of ever meeting her doctor, the woman refused to approach to shake his hand. Future prick avoided.

Episodic memories, by contrast to such associative learning, are consciously recalled and involve a virtual journey into the past. We can vividly reexperience someone hurting us, and we can take additional information into account when interpreting this recollection. If your doctor pricked you with a needle because, as you remember, he was taking a blood sample, your memory of the pain will probably not stop you from approaching the doctor in the future. Our mental time machines allow us to integrate information from different past events, and even to revisit them repeatedly in search of new lessons.[23] While this capacity can sometimes manifest in excessive rumination on a past mistake or faux pas, it can also help us understand why the present has turned out the way it has. Like Sherlock Holmes figuring out a crime, we can piece together different strands of evidence to reconstruct what most likely occurred in the past, though even here the major advantage lies in lessons for the future. Fool me once, shame on you. Fool me twice, shame on me.

Humans score a big payoff not only from pondering what has already happened but also by transplanting our memories into the future and anticipating what could happen. If the last time you went to the lake you were bitten by mosquitoes, you can project that memory forward and imagine getting bitten on your next excursion. So you may put on repellent beforehand. Our treasured capacity for episodic memory may have even evolved primarily as a side effect of this arguably much more consequential capacity for episodic foresight.[24]

Is episodic memory just the flip side of a virtual time machine that has evolved to enable us to visit the future? Does it really make sense to take the metaphor of a virtual time machine seriously enough to consider memory and foresight as complementary aspects of one neurocognitive system? After all, there are obviously fundamental differences: one deals with events that have actually happened, and the other deals with events that have not (and might never). If you want to keep track of what has been accomplished and what still needs to be done, you better not mistake one for the other.

Nonetheless, there are plenty of reasons to think that mental time travel into past and future are indeed closely related. Clive Wearing lost not only his ability to recall ever having given a performance but also his ability to travel mentally into the future and imagine giving a performance next week—or at any future time for that matter. Clive is in no position to plan for the weekend, look forward to a holiday, or plot a surprise birthday party.

Research has uncovered a wealth of other commonalities between mental time travel into the past and into the future. In one of our studies we asked young children questions about their past and future and found a link between their ability to accurately recall what they did yesterday and to report what they would do tomorrow. Similarly, as people age, the richness and accuracy of memory and foresight appear to decline in parallel. Certain clinical conditions, such as depression and schizophrenia, are associated with reduced detail during recollection, just as they are associated with reduced detail during anticipation. Furthermore, both episodic memory and episodic foresight become less vivid the more distant the events are from the present. And, as we will see later, both involve similar patterns of brain activation. These commonalities are no coincidence. We do appear to use the same mental time machine to go both backward and forward.[25]

—————

While an uncritical animal may be eliminated together with its dogmatic hypotheses, we may formulate our hypotheses and criticize them. Let our conjectures, our theories die in our stead!
—Karl Popper (1978)

It is not always enough to simply project the past into the future and assume the same thing you remember will happen again. Because the future is uncertain, imagining multiple possibilities is fundamental to effective foresight. You can consider different ways in which a friend might react to her surprise party, even if you have never seen her in such a situation.[26] What would your friend think if you persuaded a famous person she admires to appear at the party? Or how well would she take it if you played a little prank and replaced the lemon icing on the birthday cake with hot mustard? Assuming you have never tasted such a concoction, you can still be pretty confident that it would taste positively awful—even if you like both mustard and cake. You can also imagine what your friend's reaction would be when she discovers that you conspired with others in this prank. There may be retaliation, after all. In fact, your more level-headed companions may advise against such shenanigans, and so you may decide to keep your mischievous streak in check.

The point is that these considerations are not simple repetitions of the past but novel possibilities that you can entertain and evaluate from the comfort of your armchair. Your private time machine gives you access to futures that have never been—indeed, you may use it to make sure they never come to be. You can bring some ideas to life and let others die in your head, as Popper remarked—without having to suffer all the real fallout from trying them out in action.

To construct multiple versions of the future, we need tremendously flexible, open-ended cognitive machinery. Together with his mentor, psychologist Michael Corballis, Thomas has proposed that much like a theater production relies on many carefully orchestrated processes to bring events to life before our eyes, many processes must operate in concert when we generate a scenario in the mental theater behind our eyes.[27]

To imagine the fun of jumping from behind the couch as your birthday friend enters the house, you must be able to disengage from whatever is happening around you in the present moment to host the mental scenario (the stage). You also need some representation of yourself and your friends (actors). You might ask yourself what you are going to do to stop Charlie from giving it away by giggling, as he has before. You might want to picture the particulars of the house and where everyone could be hiding (the set) as well as the logistical problems this may present. Now, when you imagine the situation, you can run through a couple of different versions. Are you just going to wait for her to come at her own leisure, or might it be better to have one of you walk her there? To conjure up these potential scenarios, you need some kind of creativity to generate the options (playwright) by combining and recombining elements into new constellations. In your mind, you may work through a range of possibilities and evaluate their appeals and risks (director). These assessments take advantage of the fact that merely thinking of events can trigger corresponding emotions—which is why it can be excruciating even to imagine a very embarrassing situation and immensely pleasing to picture a victory. But although going through scenarios can be fun and entertaining, time is limited. At some point, you have to make a call and decide what to pursue in reality (executive producer).

This is not to say there is a little theater in your brain or little brain-people fulfilling these roles. The metaphor is rather meant to highlight that many components must work together to construct mental scenarios removed from the here and now. Our mental time travel is based on the interconnected capacities to imagine things not currently perceived, to entertain the actions and reactions of self and others, to understand physical and social dynamics, to generate storylines by combining basic elements, to reflect upon and evaluate possibilities, and to put a plan into action.

The human mental time machine is a complex and powerful device that allows us to imagine virtually whatever, wherever, whenever. But when it is used to travel to our past, its flexibility and open-endedness can lead to unreliable versions of what actually happened. We may incorporate later information and reconstruct events creatively rather than faithfully. The biases and errors of our memory are the price we pay for a system flexible enough to let us entertain endless possibilities.

———

A Prudent Man will look as far as he can, and provide
to the utmost of his Knowledge and Power, but when
that's done, he knows he's but a Man and therefore
can't possibly Forsee and Remedy all things.

—Mary Astell (1697)

What do you have in your bags and pockets, and what do you really need for the day ahead? Perhaps you carry cash, credit cards, chewing gum, cosmetics, car keys, contraceptives . . . to name just some common items starting with C. You cart such objects around presumably because you reasoned that at some future point they might come in handy—even if sometimes the use-by date passes first.[28] Overpreparation is characteristic of humans because we know the future is uncertain.

Thinking about the future is clearly powerful. But thinking about our thinking about the future can be even more so.[29] You appreciate that the future you imagine is *just your imagination*: it may not become reality. You can think about how your predictions might be woefully off target and how your best-laid plans often go awry. And so you can set out to compensate for these shortcomings. As the young children in our forked-tube experiments demonstrated, by preparing for more than just one possibility you can increase your chances of success. You can also think about futures where you succeed and futures where you fail—and how those alternative futures might be brought about. How do you tip fortune in your favor?

Anticipating that we might not remember what we have to do on particular days or at particular times, we use lists, calendars, and alarms. Knowing our best intentions for self-control are no guarantee, we hide our cookies, throw out our cigarettes, and transfer our money into savings accounts. Even before humans were building machines like the Antikythera mechanism to help them predict and coordinate, they reflected on future challenges and devised ways to compensate for their limitations. Foreseeing that they might not be able to plot their way back home, people sketched lines in the sand to plan a route and memorized stories about notable landmarks. Predicting that they might not have the skills they would need, they deliberately practiced to be better prepared. Realizing that they might lose track of who owed what to whom, they developed accounting systems to do the work for them. Across the board, they also used social means to overcome their future shortcomings, discussing their plans, seeking advice, asking to be reminded, or letting wiser people lead the way forward.

People don't just predict and make plans. A key to human success is that we reflect on whether our predictions will turn out to be right, and whether our plans will become reality. When the answer to either question is *perhaps not*, we seek workarounds. We prepare for the worst, even if we hope for the best. Much of our strength derives from compensatory strategies we have created to deal with the unknowns on the cloudy horizon.

———

Let's cast our minds forward to what lies ahead in this book.

In the next chapter, we take as our starting point the role of foresight in the evolution of cultural solutions to life's challenges—how to build boats that float, paint art that lasts, and cook poisonous fish so it won't kill you to eat it. We will see how foresight drives cultural evolution in a feedback loop that sets us en route to our current position on the planet.

We then consider how foresight develops and how it works.[*30] In Chapter 3, we will see how children acquire their mental time machines, and witness the powers this unlocks. In turn, we will argue that the different choices people make as they steer these time machines into the future underwrite the diverse skills and knowledge that characterize human societies. Chapter 4 then takes us under the hood, in an effort to understand how the brain enables us to think ahead and reflect on what the future might hold. Prediction is central to how all animal brains process information, and the human brain has taken this to another level, creating scenarios far removed in time and space.

Having introduced what science has discovered about the development and mechanisms of our mental time machine, we then turn to ultimate questions about the nature and origins of foresight. We start with the capacities of other animals before putting the machine into gear as we move forward through human evolution, recorded history, up to the present, and into the future.

In our own research we have studied the cognitive capacities of apes, monkeys, crows, and other animals. In Chapter 5, we will discuss what such studies have revealed about animal foresight, and what makes human minds so uniquely successful at thinking ahead. Chapter 6 takes us on a journey deep back in time, where we will meet the various other upright walking hominin species with whom our forebears once shared the planet. We will examine the archaeological evidence for the expansion of their time horizons and discover how our ancestors began their quest for mastery of the future.

It turns out that much of that mastery comes from outside our skulls. In Chapter 7, we take stock of cultural artifacts that have drastically enhanced our foresight. The Antikythera mechanism is but one of the tools—calendars, clocks, cuneiform, and other creations—that helped humans predict the future and deliberately coordinate their actions like no other

---

* The ethologist Niko Tinbergen highlighted the importance of distinguishing between two proximate explanations of a behavior (how it develops and what mechanisms are involved) and two ultimate explanations (how the behavior evolved and what functions it serves). Fully understanding the nature of foresight will likewise require considering all four questions.

creatures ever could. In the last chapter, we will see how humans harnessed foresight to bring about the modern era with all its comforts and predicaments. There are still plenty of lessons to be learned about piloting our mental time machines, and they'd better be learned quickly.

———

After hundreds of millennia living in small clans armed with stone tools, in the last ten thousand years humans have gone from turning silica stone into axes to turning it into computer chips. With better foresight and coordination, we have created more and more machines, artifacts, and gadgets that are all around us, some of which are even speeding into interstellar space.[31]

This trend has been accelerating since humans discovered the scientific method: a systematic way to build knowledge with foresight at its core. Experiments and observations give rise to theories, which lead to predictions that are then tested with further experiments and observations. If the predictions turn out to be wrong, scientists try to devise a better theory to explain the unexpected observations, which then leads to new predictions and tests. And so on. With this simple cycle—essentially an error-correction mechanism—the collaborative scientific endeavor has resulted in giant strides in our understanding of the world. In turn, it has made people ever more effective at forecasting and shaping the future.

Science has given us a new view of our place in the really big picture of time and space. Just consider that the daylight you see left the sun's surface about eight minutes before it struck your retina. And when we look at the brightest star in the night sky, Sirius, we see light that left that star over eight and a half years ago. Light from the center of the Milky Way must travel more than 25,000 years to reach us, and when you spy through a telescope on the next closest galaxy, Andromeda, you see light that originated some 2.5 million years ago. Now, if there is intelligent life in that galaxy that happens to be looking back at us, it would see the Earth as it was 2.5 million years ago, populated by multiple species of our long gone relatives— *Australopithecines*, *Paranthropus*, and *Homo habilis*—before anyone had ever left Africa. Though these bigger pictures may make us feel tiny and insignificant, they have also yielded a clearer sense of our species' epic journey. We are the last of many upright-walking hominin species that once roamed the Earth, and we have come a long way.

Partly as a result of farsighted efforts to make the world a better place, many of us now enjoy comforts such as motorized transportation and telecommunication that our great-grandparents, let alone prehistoric hominins, could not have even dreamed up. The ebb and flow of the tides is no longer an unpredictable process but a well-understood pattern that seafarers consider to avoid running their ships into the ground. Even a tsunami is no longer a smite from the gods but a predictable consequence of a geological event that early-warning systems can detect, buying people precious minutes to seek higher ground.

We can also recognize that many apparent human advances come with not-so-harmless consequences: forests are burning, glaciers are melting, and a disturbing number of species are dying out. We are extracting what we want from the planet and leaving mountains of trash in return. Our litter can be found in the deepest sea trenches and the outer reaches of the atmosphere. Human impact on the planet has increased so dramatically that scientists have declared a new geological epoch: the Anthropocene.[32] Nuclear weapons testing in the 1940s and 1950s left streaks of radioactive elements in rock layers across the globe that mark the onset of this era.*[33] Many scientific predictions about pollution, climate change, and mass extinctions suggest that we are now standing at a crossroads. It is high time to find out more about the very capacity to think ahead that has driven us there. After all, it may also be the only way out of our troubles.

Thinking ahead permeates most of our actions and is essential to human affairs. This is not a new insight. In ancient Greek mythology, Prometheus, son of the Titan Iapetus, stole fire from heaven to give humans the powers that would distinguish them from other animals. He brought us culture, farming, mathematics, medicine, technology, and writing. *Prometheus* means "foresight."

---

* Nuclear weapons testing also left high amounts of carbon-14 in animal body parts. Curiously, this helped scientists to look back in time. Greenland sharks, for example, can be assessed as having been born before or after the nuclear era. It was found that even relatively small sharks were several decades old—which enabled researchers to extrapolate the likely age of one large shark to an astonishing 392 years.

CHAPTER 2

CREATING THE FUTURE

The best way to predict the future is to invent it.
—Alan Kay (1971)

In 1812, the novelist Fanny Burney wrote a letter to her sister Esther in which she recounted a mastectomy she had undergone the year before. Her courageous report makes for grim reading:

> When the dreadful steel was plunged into the breast—cutting through veins—arteries—flesh—nerves—I needed no injunctions not to restrain my cries. I began a scream that lasted unintermittingly during the whole time of the incision—and I almost marvel that it rings not in my ears still! So excruciating was the agony. When the wound was made, and the instrument was withdrawn, the pain seemed undiminished, for the air that suddenly rushed into those delicate parts felt like a mass of minute but sharp and forked poniards, that were tearing the edges of the wound.

Burney's operation had taken place before the introduction of anesthetics into medicine. Her full account of the awful affair includes testaments to the costs of mental time travel: the many months of anxiety leading up to

23

the surgery and its long-term psychological aftereffects. She could not revisit the event in memory without immense suffering: "My dearest Esther, not for days, not for Weeks, but for Months I could not speak of this terrible business without nearly again going through it! I could not think of it with impunity! I was sick, I was disordered by a single question—even now, 9 months after it is over, I have a headache from going on with the account!"[1]

For thousands of years there was no alternative to the agony of such an operation. Alcohol and opium were about the best you could take to dull the sensations while your body was sliced open. But in the late eighteenth century, British chemist Humphry Davy made an astonishing breakthrough during a series of experiments conducted on himself. Davy would heat ammonium nitrate crystals, collect the resulting nitrous oxide gas, and then inhale it. The experience was rather pleasant, and so he investigated more often and with stronger doses. Davy had encountered the effects of laughing gas, and he invited family and friends, including the famed poet Samuel Taylor Coleridge, to partake.

In a monograph from 1800 titled *Researches, Chemical and Philosophical*, the twenty-one-year-old Davy presented the methods and results from his experiments into the production and inhalation of nitrous oxide. He reflected on the intense psychedelic effects of high doses, which amused and thrilled him and his companions. Buried amid 580 pages of dense technical details and philosophical musings was an offhand remark: "As nitrous oxide . . . appears capable of destroying physical pain, it may probably be used with advantage during surgical operations."[2]

For decades, this prophetic suggestion fell on deaf ears—and Fanny Burney still had to agonize through her operation.[3] But then, in 1844, the dentist Horace Wells, attending an exhibition of nitrous oxide in Hartford, Connecticut, noted that a man under the influence injured his leg running into wooden benches, apparently without feeling pain. Wells bravely had a molar extracted the very next morning to verify the analgesic effects of the drug, and then set out to convince the medical profession of its promise. Eventually the revolutionary potential of anesthetics became widely recognized, methods were refined, and much terrible suffering—not least when undergoing a mastectomy—was prevented.[4] Today, anesthesia is so routine we take it for granted, having found ourselves the beneficiaries of farsighted forebears passing on lessons for the future.

For millennia, humans have explored the world with curiosity and recognized potential in the things they discovered. In the process, they have gradually accumulated solutions to life's problems. They turned raw materials into endless varieties of desired products and handed down knowledge, skills, and practices to the next generation. Through their cultural heritage, human groups could thrive in diverse habitats—even in the bitter cold of Siberia, the sunbaked heat of Central Australian deserts, and now in metropolises built from steel and concrete. In this chapter, we will see how foresight has made human culture such a transformative force on our planet.

———

All animals exploit particular ecological niches, often with complex adaptations that other species lack. Giraffes reach higher, cheetahs run faster, and sperm whales dive deeper. Koalas subsist on eucalyptus leaves, and dung beetles can digest the already digested. Human uniqueness is no different. Our ancestors specialized in exploiting what evolutionary anthropologists John Tooby and Irven DeVore called the *cognitive niche*.[5] Early humans increasingly predicted the goings-on in their environments—predators and prey, seasons and storms—until eventually they plotted their way to dominance wherever they went. What's more, they began to construct artificial environments that catered not just to present but also to future needs. Why accommodate to your world when you can change the world to better accommodate you?

Foresight allowed us to keep the carnivores at bay and rise to the top of the food chain. It is probably no coincidence that most of the Earth's megafauna, from mastodons to moas, routinely went extinct as humans migrated from continent to continent and rowed from island to island.[6] Lands once dominated by such enormous beasts quickly fell under the rule of cunning primates who, on the face of it, should have been well out of their depth. We had the opportunity to conquer because of our smarts, coordination, and farsightedness. And conquer we have. Our weapons—pitchforks, poisons, projectiles—can destroy any biological enemy. Today humans dominate virtually the entire globe, and we have given ourselves the fitting name *sapiens*: the wise.

If this all sounds a bit hubristic to you, then you are not alone. Are humans really a bunch of brainiacs who mastered the natural world with sophisticated stratagems and clever reasoning? Critics have been quick to point out that this perspective turns a blind eye to the *sapiens'* track record of stupidity and, more importantly, to the enormous dependence of individuals on the hard-won and gradually accumulated cultural practices of groups.

In his book *The Secret of Our Success*, anthropologist Joseph Henrich illustrates this point with a series of tales from the age of European exploration.[7] Many explorers got lost, shipwrecked, or otherwise abandoned in the wilderness during this time. Destitute parties often dwindled in number as man after man died at the unforgiving hands of Mother Nature. Despite the struggles of the newcomers, indigenous inhabitants were not just surviving but often flourishing in those very same ecosystems. The newcomers struggled because they did not have the cultural knowledge and skills the locals had accrued over thousands of years—including how to turn poisonous plants into edible foodstuffs, to find waterholes, and to exploit available varieties of flax, bark, furs, tubers, or flint.

According to Henrich and other scholars, "cultural evolution" can be thought of as the process by which profitable variants of ideas, techniques, and tools accumulate, and so create sophisticated solutions to even tricky problems.[8] No individual person figured out all the clever behavioral techniques needed to thrive in lands where the lost explorers struggled; instead, these techniques emerged over many generations. One individual copied someone else, perhaps making small changes, before being copied in turn. People who inherited the routines in this way could follow them blindly even if they did not fully understand why they were filling out each step. By the same token, many of the technologies we benefit from today were incrementally acquired rather than devised wholesale by farsighted geniuses.[9] The surprising conclusion from this view, then, is that many cultural artifacts and practices only *seem* to have been designed by deliberate human reasoning.

In short—so the argument goes—humans really aren't as smart as it seems at first blush. The secret to our success is cultural evolution. And it may not have required much intelligence or foresight at all; just like biological evolution can lead to sophisticated adaptations without anyone planning them.

Let's take a brief detour to consider what it means to say biological evolution lacks foresight. Writing an account of his life for his family to read after his death, Charles Darwin reflected on the implications of his evolutionary theory for religious belief. Did his explanation for the origin of species do away with the guiding hand of God? Darwin was long troubled by such questions. He was a great admirer of the clergyman and author William Paley, who had argued that the complexity of living things was irrefutable evidence of a divine designer. By the time Darwin was penning his autobiography, however, it was clear to him that the theory of natural selection had conclusively replaced the argument for design: "We can no longer argue that, for instance, the beautiful hinge of a bivalve shell must have been made by an intelligent being, like the hinge of a door by man. There seems to be no more design in the variability of organic beings and in the action of natural selection, than in the course which the wind blows. Everything in nature is the result of fixed laws."[10] It is now accepted in biology that the apparent design of living beings is an illusion. Instead, we know that adaptations, by which organisms appear to neatly fit their environments, result from a gradual process of mutation and selection. Individual organisms differ from each other, and more individuals are born than can survive. Some are more likely to reproduce than others, and these pass on their lucky variations. The next generation therefore includes more copies of these variations than the previous generation. Over time, organisms function increasingly well in their environments. Over vast spans of time, especially if there is some geographic isolation, this descent with modification produces new species. This is Darwinian evolution by natural selection in a nutshell. It completely lacks foresight, as famously highlighted by Richard Dawkins in his book *The Blind Watchmaker*.[11]

So in Henrich's view, culture "evolves" similarly by the accumulation of minor changes rather than through clever foresighted human reasoning.

And, ultimately, the capacity for cultural learning itself must have evolved via natural selection. The roots of culture are biological. One possibility is that humans are even born with an ability to copy their parents, at the ready to acquire their cultural inheritance.

————

A baby gazes up at her mother. The mother, cooing, sticks out her tongue at her child. Lo and behold, out comes the infant's tongue. Beginning in the 1970s, a series of famous experiments reported that even newborn babies can imitate facial gestures, including sticking out the tongue, opening the mouth, and pouting.[12] This was a remarkable finding, not least because babies can only see other people's facial expressions and not their own. *Neonatal imitation* was hailed as the foundation of human social cognition, even though it remained mysterious how a baby fresh from the dark confines of the womb could copy at all. In any case, these results suggested humans are innately prepared to copy the behaviors of those around them and so absorb their culture.

In 2016, working with developmental psychologists including Janine Oostenbroek and Virginia Slaughter, our group published the largest-yet longitudinal study of neonatal imitation.[13] We had intended to track how differences in infant imitation might relate to later social learning and cognitive developments. But to our surprise, the babies did not copy any of nine actions at any of our initial four testing occasions between birth and nine weeks of age. Yes, sometimes babies poked out their tongues after an adult did so, but they were just as likely to do this in response to other facial expressions, such as smiling or frowning.

For forty years, parents had been told that their newborns could imitate their gestures; many even worried when their little ones didn't copy along. Our data challenged the very existence of the phenomenon whose trajectory and consequences we had sought to chart—and further research has bolstered this challenge.[14] The mystery of how newborn babies could possibly copy what another individual does has been solved. They don't.

The psychologist Cecilia Heyes declared that our findings showed imitation is learned rather than innate.[15] From Heyes's perspective, human imitation is not particularly special, but is instead acquired through the same old associative learning mechanisms we mentioned in the previous chapter. For instance, because people often copy their infants, babies may gradually learn to associate what they do with what they see. A mother might start by poking her tongue out whenever her baby does, but eventually the baby links these actions and begins to produce tongue pokes in response to the visual input of the mother's tongue poke.

Associative learning is everywhere in the animal kingdom. So if associative learning really is all there is to imitation, and imitation opens up the possibility of cultural evolution, one might expect apes and other social creatures to also evolve cultures. In 1999, the prestigious journal *Nature* featured an article titled "Cultures in Chimpanzees."[16] Evolutionary psychologist Andrew Whiten had recruited the leaders of seven long-term field studies of chimpanzees in the wilds of Africa—including famed primatologists Jane Goodall, Richard Wrangham, and Christophe Boesch—to systematically compare notes. The team realized that certain behaviors were common in some communities but totally absent in others, even when there were no obvious ecological explanations. At Gombe in Tanzania, chimpanzees would insert long sticks into ant nests and wipe the ants off with their fingers, while chimpanzees in the Tai forest of the Ivory Coast were eating their ants straight off a much shorter stick. Chimpanzees at Mahale would frequently groom each other while holding hands in the air, as if giving each other a high five. By contrast, the chimpanzees at Gombe, only around ninety-five miles away, tended not to hold hands while grooming. The Tai forest chimpanzees were using wooden and stone hammers to crack nuts, whereas those at Bossou in Guinea were using only stone hammers. And at Gombe they were using neither.

Experimental studies since Whiten's paper have confirmed that chimpanzees can learn from each other how to solve problems. The evidence is compelling: dozens of behavioral differences between groups of chimpanzees are the result of socially maintained traditions.[17] An archaeological site in the Tai forest studied by Boesch and his team even suggests that the chimpanzees there have been passing on their nut-cracking technique for over 4,300 years.[18]

Scientists have now observed several other species transmitting behaviors socially, and some researchers suspect this is just the tip of the iceberg.[19] Nonetheless, the number of socially maintained traits in other animals pales in comparison to the many thousands of idiosyncrasies found in every human group: burial rites, recipes, coming-of-age rituals, melodies, proverbs, bodily adornments, customary greetings, jokes, taboos, counting systems, moral rules, and hairstyles. So what is it that makes human cultures so much richer—and more powerful—than those of other animals?

Though we might not be born with the capacity to imitate, there is, in fact, something quite peculiar about how human children imitate that seems to have no parallel in the animal kingdom. In a series of studies, Whiten and his colleague Victoria Horner presented a puzzle box to young children and young chimpanzees.[20] An experimenter showed them how to get the treat inside: first, poke a stick into a hole at the top, then insert that stick in another hole on the side. Both children and chimpanzees readily copied these simple actions to get the treat. In a second condition, however, the puzzle box was made out of transparent plexiglass. Now it was clear that the first of the two actions was completely superfluous. To open the box, one merely had to poke the stick into the hole at the side. Chimpanzees swiftly recognized this and ignored the experimenter's first action. They poked the stick straight in the side. The human children, by contrast, continued to imitate the pointless poke into the top of the box. The young chimpanzees, in other words, did much better than human children; at least when you consider who learned the more efficient way to open the box.

This experiment reflects something important about human imitation: children tend to faithfully copy what other people are doing, regardless of whether they understand why the models are doing it. Our penchant to *overimitate* ensures the faithful transmission of hard-learned lessons, even when ignorant youngsters do not yet fully grasp their future utility. It allows culture to evolve effectively because solutions to problems can be passed on from one individual to the next without requiring each person to figure everything out from scratch.

Cultural evolution researcher Maxime Derex and colleagues tested whether adults could gradually solve a problem through social learning without anyone really understanding how the solution worked—in line with Henrich's view of cultural evolution. The problem was to get a wheel to roll down a track as fast as possible by choosing where to place weights on its spokes. This might sound like a walk in the park, but wheel dynamics are not entirely obvious. Participants were put into chains of five. Working in isolation, each person was required to make as much progress as possible, and their final attempts were shown to the next person in the chain—the next generation, as it were. Over the generations, people did in fact get the wheel cruising down the rails faster and faster. But when they afterwards had to choose between sets of two different wheels, those in generation five were no better than those in generation one at picking the winners. The experimenters concluded that, mentally, the gears weren't really turning. The participants' progress in getting the wheels to move faster did not derive from an increasing understanding of what was going on.[21]

Yet participants did not place the weights on the wheel at random. For instance, they were more likely to correctly shift the center of mass to the top of the wheel than to the bottom. As the experimenters acknowledge, people came to the problem with some idea about how to solve it.[22] Human cultural evolution is not just an extended process of trial-and-error learning and (over)imitation, spread out over generations. People model the relationships between phenomena in the world around them and anticipate solutions.

Anyone who has spent a bit of time in the kitchen, for example, will have followed time-tested, traditional recipes but also spiced things up with new ingredients. We do inherit cultural knowledge about certain tasty combinations—*recipe* derives from the Latin *recipere*, "to receive," after all—but we also draw upon our own personal history of gustatory experiences. These experiences are food for thought; we combine ingredients in our minds before ever putting them into the pan. Tuna? Tasty. Strawberry milkshake? Delicious. What about tuna in your strawberry milkshake? You be the judge. The novel combinations that we actually try out are not random, nor are they based on a simple good + good = good formula. Instead, they are based on our capacity to preexperience novel constellations of more basic sensations. If we happen to hit on a particularly delicious combination, it might become part of the cultural tradition. Grandma's recipe lives on.[23]

The current trend of emphasizing mindless cultural evolution runs the risk of selling our smarts short. Yes, humans rely on socially maintained traditions even more than chimpanzees do, and our tendency to overimitate may be one reason we accumulate traits more effectively. But humans also exploit the cognitive niche. There are two critical ways in which foresight makes a difference to how our cultures evolve: instructing and innovating.[24]

Let's tackle each in turn. By the end of this chapter, we hope to have shown how foresight and cultural evolution are inexorably linked in a feedback loop that set humans on the path to the Anthropocene.

Imitation need not require much, if any, awareness about the benefits of learning or about the prospects of the remote future. In fact, as the studies above demonstrate, imitators often simply follow along, even if they are oblivious to the reasons for their actions. Yet, humans have another way of transmitting culture that is, at its core, reliant on the foresight of clued-up individuals who actively pass on what they think others will need. We transmit many cultural traits deliberately. We teach.

People understand that youngsters are ignorant and recognize that learning in the present will benefit them in the future. Parents regularly intervene in their children's activities to teach them when the opportunity presents itself. *Look both ways before you cross the road.* Not only do we stop the little ones from getting hurt, but we also highlight the lessons they ought to absorb. *Expect drivers not to notice you.* With language we can point out the important cues to which learners should attend and explain the actions they should take. A teacher may also act out the desired sequence, perhaps exaggerating the most significant actions, and encourage imitation. *Watch me first! Now you do it.* Teaching turns imitation into an even more potent tool that can be wielded strategically.

The most famous example of apparent teaching in animals comes from meerkats, where the young must learn how to deal with venomous scorpions. The adults first introduce dead scorpions to the young. As the pups get older, the adults begin to bring live scorpions, but first they remove the sting. Eventually, adults bring intact prey for their pups to handle. The gradual exposure protects the young as they learn to deal with such danger-ous quarry. However, although this looks like clever teaching, it turns out that the adults merely respond in a fixed manner to particular stimuli. If researchers play the begging calls of old pups, the adults will bring back live scorpions even if their actual pups are too young to deal with them safely.[25] Animals can certainly act in ways that function like teaching, without any need to propose that they understand what they are doing. Indeed, when one simply defines *teaching* as individuals modifying their behavior in a way that fosters learning in a pupil, then meerkats, cats, and even ants meet the criteria.[26] But there is little to suggest that other animals think ahead to what another individual needs to know in order to deliberately teach them.

There are some clues about the origins of deliberate teaching in the ar-chaeological record. As we will see in Chapter 6, when we discuss the evo-lutionary origins of foresight, versatile bifacial handaxes are evident from around 1.8 million years ago.[27] In a recent study, archaeologists taught mod-ern humans the techniques to make these tools.[28] Although the students received some eighty hours of training and improved over time, they still did not achieve the expertise required to make handaxes as sleek as the ones frequently found in the archaeological record. The ancient stone knappers must have been committed learners, and possibly dedicated teachers. Other studies in which students learn stone tool knapping from each other have found that active teaching—and especially verbal instruction—improves the transmission of the skill. While some researchers have therefore con-cluded that "teaching or proto-language may have been pre-requisites" for the appearance of this stone tool technology, that may be pushing the en-velope a little, given that the participants have grown up with language and would be expected to benefit most from verbal instruction.[29] Unfortunately, we cannot test the original makers of the tools: *Homo erectus*.

Still, it is uncontroversial that today, verbal instructions are immensely powerful in transmitting skills and knowledge. In one study investigating the spread of ideas during problem-solving, even three- and four-year-old chil-dren readily taught each other how to retrieve a prize from a puzzle box, while chimpanzees and capuchin monkeys did not.[30] And children performed far better when they received verbal guidance on how to achieve the goal. While intuitive, these results support the suspicion that verbal teaching, whenever it first evolved, must have dramatically accelerated cultural evolution.

With language, people can not only teach each other skills without having to play charades but also transmit insights, plans, and predictions. Even just recounting what happened to you yesterday can be instructive. If you found out the hard way that swimming in the big brown lake is dangerous because you were chased off by a crocodile, you can then pass on that information to friends before they plan their own sunset dip. This advice can be proffered back at the campsite because mental time travel lets us revisit and convey events we have experienced even long after we have dried off.[*31]

With or without language, teaching is not a blind process. It frequently involves peering into the future to strategically pass on the right information in the right way at the right time. It also often exploits the much older mechanism of imitation to great effect. *Copy this.* But where do new ideas come from? In biological evolution, random mutations in DNA molecules are the source of new variation, and natural selection retains the beneficial variants. Mutations to ideas typically come about differently. This brings us to the second major aspect of cultural evolution that depends on foresight: innovation.

————

Before Thomas could start elementary school in Germany, he remembers taking a test in a big hall together with several other children. One of the tasks involved a rubber duck, which was placed on the floor. Thomas was given a long, straight wire and told to use it to pick up the duck. There was no way to achieve this with the wire, at least not before he glimpsed a neighbor picking up the toy with a hooked wire. Little Thomas had the solution; he bent the tool and retrieved the duck, relieved to have figured it out, and

---

* In some cases, oral traditions are said to reach back many generations. An Australian Aboriginal story about the formation of Lake Eacham near Cairns, for instance, recounts how the sky turned an unusual red after some youths violated a taboo. The earth roared like thunder, then cracked open, swallowing people and creating the lake. Analyses of the sediment in the lake suggest it was formed over nine thousand years ago in a violent volcanic eruption.

yet feeling guilty for having copied. Upon hearing this story, Thomas's ten-year-old son reckoned the professor should return his PhD because clearly he had cheated on the primary school entrance exam. Thomas's parents, on the other hand, thought this problem-solving demonstrated he was ready to go to school.

The ability to solve this kind of task is not uniquely human. When two New Caledonian crows were presented with a narrow, transparent tube containing a small bucket of food and two pieces of wire—one straight and the other hooked—the crows worked out how to use the hooked wire to raise the bucket.[32] A few trials into the study, however, one crow removed the hooked wire from the cage. The other crow then spontaneously bent the straight wire into a hooked shape and used it to retrieve the bucket from the tube. In follow-up trials this clever crow repeatedly bent straight wires into hooks. This report is often presented as one of the most compelling cases of animal innovation. Yet, much like Thomas at the entry test, the crow had seen that a hooked wire could do the trick. Indeed, the bird had already used one. Though impressive, and Thomas's parents may well have said the crow was ready for elementary school, it did not innovate something new.

Humans frequently conjure up their own new solutions to everyday problems. Of course, few of us are geniuses like Mesopotamian engineer Ibn al-Razzaz al-Jazari, who over eight hundred years ago invented a hydropower supply system, a series of accurate and elaborate water clocks, and a humanoid robot that would emerge from behind an automatic door eight times an hour to serve wine to party guests—but we all have the basic capacity to create something new and useful. In kitchens and workshops, music studios and sports complexes the world over, ordinary people do not just copy along but also come up with clever new ways of doing things. We can imagine the functions and aesthetics of the things we might make before investing in the real deal, and so we try out small-scale versions in our minds, on paper, and in models. From an individual imagining how to reshape a coat hanger to unclog a drain to an R&D department of a plumbing company testing different options in search of a better unclogging device, our problem-solving often critically depends on foresight.[33]

In fact, innovation depends on foresight in a most fundamental way: it enables us to recognize future utility.[34] Without this recognition, even the best, most creative solutions may be ignored or tossed aside once the present problem has been dealt with. By appreciating the potential of objects and ideas, people become motivated to retain them and to refine them; to share, sell, or steal them.

Recall that the discovery of the sensation-dulling properties of nitrous oxide was followed by decades of people using the gas for recreation before it was eventually deployed as an anesthetic. History is replete with what appear to be baffling examples of missed opportunities for innovation. In pre-Columbian Mexico, somebody invented the wheel. The wheel has long been considered a pinnacle of human achievement, as it unlocks a great variety of valuable tools, from pottery to pulleys, and from carts to clockwork. Yet here the wheel appears to have been deployed solely for the purpose of children's toys: small animal figurines that roll along the ground.[35] Likewise, the Greek mathematician Hero of Alexandria described a steam engine in the first century AD but seems to have used it only to entertain guests at parties.[36] If Hero or one of his guests had recognized the myriad practical uses of such a device, then—for better or worse—the Industrial Revolution just might have kicked off many centuries before it actually did.

As long as someone recognizes the future utility of a solution, it does not really matter whether they creatively came up with that solution or stumbled on it serendipitously. When Alexander Fleming went on vacation in 1928, he had no idea that upon returning to work he would have the opportunity to change the course of medicine forever. But while he was away, some contaminant mold just happened to grow on one of his Petri dishes, killing the surrounding staphylococci bacteria.[37] Fleming recognized the potential of the *Penicillium*, and so he started examining which other harmful bacteria might be killed by it too. It was little more than a decade later that the first person with a life-threatening infection was cured by penicillin, following extensive work on turning the fortuitous discovery into a purified drug.*[38]

---

* Occasionally we stumble upon something useful while pursuing a completely different objective. In 1969, employees at the company Bostik were attempting to create a new sealant by combining oil with rubber and chalk powder. The resulting lump of putty was useless as

Figure 2.1. Hero's early steam engine. The many practical uses of the idea to use steam to generate power, which in hindsight may seem obvious to us, were apparently not recognized by its creator.

As tales of accidental discoveries and missed opportunities illustrate, recognizing future utility ladder is critical to innovation.

With foresight we can imagine solutions, recognize potential, recruit others in our quests, and teach them what we have learned. We can even recognize the potential of an idea without knowing how it could ever really work. When science fiction writers such as Jules Verne envision light propelled spacecraft, submarines going twenty thousand leagues under the sea, or flights to the moon, other people can adopt these visions as goals and start working on how to realize them. Many lofty ideas have eventually become

---

a sealant, but some employees allegedly started taking chunks of it from a tray on someone's desk to stick up messages around the office. The higher-ups caught on to this unexpected utility of the putty, and it was branded with enormous success as Blu Tack.

reality, even if others, such as Verne's notion of a journey to the center of the Earth, have not.

————

So how do we create the future? What explains the apparent ecological dominance of *Homo sapiens*? How did our species swell, in the span of just a few hundred thousand years, from scattered bands of hunter-gatherers eking out a living on the African savannah to several billion individuals distributed across most ecosystems on the planet, and even in orbit around it?

Many scholars have answered the question about human ecological dominance with an enthusiastic embrace of cultural evolution. We have seen in this chapter how the transmission of ideas from mind to mind within and between generations leads to the accumulation and refinement of solutions to problems. Through this second inheritance system—a system that other animals have only evolved to exploit in a minuscule way in comparison—humans created better and better solutions to life's challenges. We also saw that children's tendency to overimitate enables the transmission of solutions even in the absence of causal understanding. Copying may have helped human societies to build up treasure troves of ideas—customs, values, philosophies, arts, tools, and buildings—from which we all benefit, just as we benefit from the inherited genes that give us eyes and ears, grasping hands, and running legs.

But humans are evidently not just receptacles and regurgitators of ideas. As cognitive scientist Steven Pinker put it, "Human cultural products are not the result of an accumulation of copying errors, but are crafted through bouts of concerted brainwork by intelligent designers."[39] We explore, investigate, and test. We plot, anticipate, and reflect. Though there are many parallels between them, cultural evolution is not blind like biological evolution is. Instead, we argue that foresight drove cultural evolution in at least two key ways:

First, through *teaching*: when those in the know convey information to those out of it, by anticipating what a pupil needs to grasp and then shaping their mind towards the goal of expertise. From making stone tools around the campfire to geology courses at university, our species has deliberately passed on skills and knowledge to each other and the next generation.

Second, through *innovation*: recognizing the future utility of solutions and propagating them. Even serendipitous discoveries, stumbled upon, are only innovations because someone saw potential and acted on it.

So foresight has clearly accelerated cultural evolution. In turn, cultural evolution leads to improved foresight, creating a potent feedback loop that explains how our ancestors carved out both a cultural and cognitive niche and paved the way to increasing human dominance on the planet—and all the new problems that entailed.[40] Let's see how foresight and culture work together with an example of the feedback loop in action.

After the last ice age, some tribes in the Levant began to abandon a hunter-gatherer life in favor of a sedentary agricultural existence that raised some novel challenges, including the collection and distribution of grains, meat, and other goods through trade and taxes. They needed a way to keep track of who owed what to whom and when it was due. An innovative solution was the use of clay tokens of different shapes, such as cones and cylinders, to stand for measures of grain or livestock. By around five thousand years ago, Sumerians began to place such tokens into hollow clay balls to record taxes that had been paid or goods to be traded. Storing information in a sealed container prevented future disputes and eliminated the need to rely on inadequate human memory. But to check their content, these balls had to be broken. Perhaps because of the wastefulness of this destruction, someone came up with a further innovation: by pressing each token on the outside of the not-yet-fired clay ball, one could create impressions of what was inside. Four cone-shaped indentations on the outside meant there were four cones on the inside. As there was no longer any real point to putting tokens inside—given that the information was now also on the outside—the balls were soon replaced by flat surfaces that made it easier to make marks in the wet clay. Mere impressions were then complemented with pictures that were traced, such as an ear of barley. For centuries, accountants invented new symbols and went busily about teaching each other how to interpret and use them.

Figure 2.2. Envelope and the tokens it held, from Susa, Iran, ca. 3300 BC. The tokens stand for commodities like animals and grain.

What we just described, of course, is how the Sumerians invented writing.[41] Clay tablets and the first scripts were initially used to keep track of economic transactions, and only a select few students attended scribal schools.[42] But as we all know, writing was eventually used to record and exchange much more, and now children the world over spend years being taught how to read and write. Writing allows us to store stories, laws, and manifestos. It turns the flow of ideas into concrete objects, freeing up mental load to reflect on and further develop thoughts. It lets us share insights and solutions; and it is, itself, a teaching tool par excellence. Writing, a product of cultural inheritance, radically enhances the very mechanisms of innovation and teaching that brought it about in the first place.

Consider what cosmologist and science communicator Carl Sagan said about how writing knits together human minds across time: "What an astonishing thing a book is. It's a flat object made from a tree with flexible parts on which are imprinted lots of funny dark squiggles. But one glance at it and you're inside the mind of another person, maybe somebody dead for thousands of years. Across the millennia, an author is speaking clearly and silently inside your head, directly to you. Writing is perhaps the greatest of human inventions, binding together people who never knew each other, citizens of distant epochs. Books break the shackles of time."

With the rise of writing, not to mention with the invention of the printing press, the rate of cultural change has increased dramatically. Today, we can find out about new discoveries almost instantly because our minds are wirelessly hooked together across time and space. The internet enables us to copy and teach like we could never do before. It satisfies our urge to connect and inform each other about virtually any idea; it lets us learn instantly from the thinkers of bygone times. If you want, you can watch Sagan speak the words you just read in the next few seconds by hopping online.

With a farsighted species at the helm, cultural evolution reached new heights. Isaac Newton famously said: "If I have seen further it is by standing on the shoulders of giants."[43] But it helps if those giants are also standing up, attempting to catch sight of something on the horizon.

Humans have created safety where we were hunted, cures where we were sick, and entertainment where we were bored. We also created weapons where we were vengeful, chopped down forests where we needed timber, and drained swamps where we wanted to farm. In the process, and with endless toil, humanity has remade the world. We are increasingly capable of shaping the future to our own design—including, as we shall see, our own future selves. In the next chapter we take a closer look at how each person acquires these powers.

CHAPTER 3

INVENT YOURSELF

The mind is at every stage a theater of simultaneous possibilities.
—William James (1890)

In the German town of Ulm, on March 14, 1879, Pauline Koch gave birth to a baby boy who did what all babies do—cry, feed, cuddle, and sleep—but who would grow up to revolutionize our understanding of time with his theories of relativity. Babies, Einstein or not, show no signs of traveling mentally in time, which should perhaps not surprise given that they can barely even travel in space. In fact, human babies must be some of the most helpless creatures in all of creation. It takes months before they can hold up their head, let alone walk, talk, and plan. Even when children are comfortable balancing on their own two feet and holding a simple conversation, parents realize their young ones have a weak grasp of the future. So it is adults who have to think ahead and pack the warm jacket, the sandwich, and the juice box. It takes time to become a time traveler.

How do we assemble our time machines and eventually seek to shape our own destiny? At the start of the book, we introduced you to a theater metaphor to illustrate how a host of component capacities work together to enable human foresight. At least some of these capacities must mature enough before the show can get on the road.[1]

Recall that mental time travelers need something like a *stage*, a means to entertain ideas while the processing of immediate reality is temporarily put on the back burner. At its extreme, this enabled sixteen-year-old Einstein to imagine what a light beam would look like if you could run alongside it (a scenario that led him to think about relativity). Among the earliest outward signs of such a mental stage can be found in the way children play. By around eighteen months, toddlers show the first bouts of pretend play—transfiguring sticks and stones into swords and phones.[2] Over the subsequent years, children spend more and more time in imaginary worlds.[3] While they traverse these simulated realities and adopt roles—mom, dad, chef, builder, doctor, explorer—they learn what to do and what to expect. Initially, children can only keep a few things in mind simultaneously, which makes for some rather restricted imaginings. A simple way to observe this shortcoming is to tell them a series of digits (4-2-7) and ask them to repeat it backward (7-2-4). As you increase the number of digits, you'll quickly see the limits of working memory.[4] The number of items children can juggle in mind grows consistently as they get older—the mental stage gradually expanding, as it were, from a soapbox to a fully fledged theater.[5]

Of course, the mental stage is typically not populated with digits but with a host of objects and actors. We know even toddlers can picture things mentally because they search for a toy once it has disappeared from view, or for a person who has left the room.[6] In one of the tasks used in our laboratory, an experimenter puts a treat in a cup and then moves that cup under two different boxes, before revealing that the cup is now empty. Children should expect the treat to be in either of those two boxes but not in any other boxes the experimenter did not visit with the cup. Twenty-four-month-olds reliably pick the logical choices, demonstrating that they can represent what might have happened to an object hidden from their view.[7] By this stage of development, events that occur out of sight are not always out of mind.[8]

To accurately represent the world in their mind's eye—the *set*—children also need some understanding of the properties of matter and motion.

When five-year-old Einstein was shown a compass by his father, he was mystified by the invisible forces that could move the needle. In general, while some basic understanding of the physical world emerges early in infancy, much of causal reasoning develops a lot later.[9] Consider the trajectory of a billiard ball when it bounces off a side cushion. Jean Piaget, the famous developmental psychologist and pioneer of the hiding experiments we just discussed, found that while five-year-olds have some basic understanding of how a ball is likely to move across a table, only by about age eight do children appreciate the predictable angles of the ball's trajectory.[10] Children can do this even before they can explicitly describe the relevant physical rules, such as that the angle of incidence equals the angle of reflection, just as adults in the previous chapter could effectively place weights on wheel spokes without fully appreciating the underlying physics. With experience and an increased understanding of such rules, the balls can be brought under a player's spell and their paths controlled.*

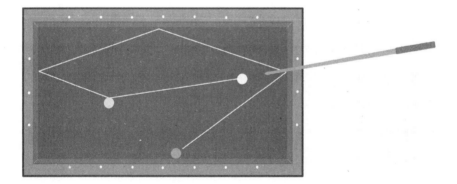

Figure 3.1. The human capacity to predict the movement of objects can eventually enable exquisite control. Consider the extraordinary trick shots that expert players can execute in the game of "three-cushion billiards," especially with a bit of added spin, to make their ball hit two other balls—but also three sides of the table.

---

* By formulating specific equations about physical laws, Einstein was able to calculate the deflection of light from another star caused by the sun's gravity. His predictions were later confirmed through measurements during a solar eclipse, as the orbs of the sun, moon, and Earth came into alignment. This finding supported his general theory of relativity and the notion that gravity is the result of a curvature of space around objects. The *Times* of London ran the following headline: "Revolution in Science. New Theory of the Universe. Newtonian Ideas Overthrown."

Children also need to gradually learn about the apparent forces that drive *actors*—intentions, desires, beliefs. Even within the first year of life, infants appear to adopt what philosopher Daniel Dennett calls "the intentional stance"—they can identify that an agent is acting as if it has a goal. Later, toddlers grasp that people differ in their tastes and desires before recognizing that they differ in what they know. Later still, around age four, children appear to appreciate that other people can have false beliefs. For instance, little Maxi understands that his mother will search for her chocolate where she falsely believes it is located, say in the cupboard where she left it, even if Maxi knows that the chocolate is no longer there (and that it tasted delicious). From this point onwards, children can deliberately tell lies—when they know something is not true but want someone else to believe that it is. An understanding of more complex interplays in the social world, such as hidden emotions, faux pas, or sarcasm, only becomes evident as children get older still.*[11]

To generate scenarios of the future, such as that surprise birthday party, children also need a capacity akin to a *playwright*. In language, we combine letters into words and words into sentences. This process also plays out across many other domains, from music to math, or in the new models that children build from the same old bucket of Lego blocks. Likewise, we can mentally combine and recombine elements such as people, objects, and actions into new patterns, and embed them into larger plots. Children typically show such combinatorial capacities from around the fourth year of life. For instance, they begin to embed sentence phrases, such as this one, within other phrases.[12] Even though human minds entertain a limited set of elements, these elements can be combined to construct a virtually endless range of new mental scenarios. Rather than random jumbles of *what, where,*

---

* Faux pas can be particularly difficult to grasp. Say Fred visits Jane, who has just moved into a new house. Fred comments on how beautiful the place is, except for those shocking curtains—not realizing that Jane just installed the curtains herself. With increasing age, children begin to wrap their heads around the higher levels of "mental nesting" involved here, meaning they can appreciate that Jane *knows* Fred did not *know* that she *thought* that the curtains were great, and that, though she *understands* that Fred *intended* no offense, she still *feels* annoyed—all the while Fred has no idea what has gone wrong.

and *when*, though, the playwright assembles these elements with narrative structure—storylines that connect events.

What's more, with the *director* at the helm, scenarios can be compared and evaluated. *Is this event likely, desirable, inevitable . . . just wishful thinking?* The inner director can adjust the details of anticipated events, assess consequences such as how other people might react, and consider alternative courses of action. As we saw in Chapter 1, by age four most children can readily consider two alternative possibilities that lie in the immediate future and prepare accordingly by covering both exits on the forked-tube task.[13] Our research group found similar competence in children from Brisbane and from two geographically isolated communities of Indigenous Australians and South Africans, suggesting that this ability to prepare for two alternative futures develops universally in early childhood. However, many other abilities associated with the director develop further along in cognitive development. By around age six, for instance, children can imagine alternative versions of past events and evaluate if these alternatives would have been better or worse than what actually happened.[14]

Eventually, like an *executive producer*, children must learn to decide what option to pursue, when to persist with their plans in the face of alternatives, and when to change tack. One aspect of this capacity for executive control is the ability to inhibit actions: to hit the brakes. Inhibition stops us from buying the first bicycle we see for sale—or stealing one that isn't—and creates the opportunity to consider how we would pay for it or what exactly we'd use it for. But inhibition takes time to emerge, as anyone who has taken a toddler to a candy store can attest. If you ask children to sort some cards based on, say, color, and then ask them to switch the rule and sort according to shape, three-year-olds struggle to prevent themselves from acting on the old instruction, whereas older children more swiftly adopt the new rule.[15]

Children also need executive control to stop pondering possibilities and put thought into action. Unlike a computer that can rapidly churn through all permutations of a problem, evaluate each solution, and then identify which one is optimal, even adults typically lack the time and resources for such an endeavor. Yes, we might want to consider all the specs and reviews of bicycles in our price range, but upon reaching some basic criteria, we tend to make a decision rather than waste more time deliberating. As playing a few rounds of Simon Says will quickly demonstrate, executive control continues to develop throughout childhood. Even the ability to follow an exceedingly simple but counterintuitive instruction (such as to look left when a light appears on the right) continues to improve well into adolescence.[16] With limited executive control, children and teenagers—even if they foresee what they need to do—may struggle to suppress immediate urges and temptations.

The components of mental time travel develop along varied and often slow trajectories. Shortcomings in any one of these can forestall a child's foresight. By the same token, considerable competences in each domain must develop before a person can have a firm grasp of tomorrow and beyond (let alone construct predictive models of the nature of time itself). In this chapter, we explore how this transformation unfolds, and see how it forges adults who can not only shape the world to their design but also set out to shape themselves. The emergence of foresight allows us to embark on journeys of further development, to deliberately change who we are.

———

Traditionally, psychology has focused much more on memory than foresight, and developmental psychology is no exception.[17] In recent years, however, the development of foresight has become an increasingly popular research topic. The simplest way to find out about children's understanding of the future is to ask them and their parents directly. After all, children also acquire something akin to a *broadcaster*—the language to communicate their mental scenarios. In one study led by Janie Busby Grant, we found that only 40 percent of parents believe that their three-year-olds often, if not always, use the word *tomorrow* correctly. This percentage gradually increases with age. While most parents of preschoolers think their kids understand words like *later*, *soon*, and *after*, other phrases pointing to more distant times—even just *next week*—appear to be mastered only when they are older.[18]

The way parents speak with their children, such as how much detail they include in conversations, predicts how much children themselves will elaborate about past or future events.[19] In dialogues, even toddlers commonly report some information about what is going to happen, and by the end of the preschool years they can talk quite clearly about various future situations—just ask them about their upcoming birthday party.[20] However, it is important to bear in mind that children may understand much more than they can articulate. And, conversely, they may say things without fully understanding them. Children may, for instance, insist that Mo has *forgotten* where the cookies are if he cannot find them, and that Anita has *remembered* where they are if she finds them—regardless of whether either of them actually knew where Grandpa put the cookies in the first place.[21]

Similarly, even though by age four many children can use the words *yesterday* and *tomorrow* correctly, research from our team has found that they tend to struggle with very basic aspects of what these terms actually imply. When children are told that Laila found out, say, the color of an obscure animal yesterday, while Kim will only find out that information on her visit to the zoo tomorrow, the children struggle to identify who of them has that knowledge *now*.[22] Such tests reveal that children may not mean the same thing as adults do when they use these words. It is therefore important to supplement insights gained through conversations with behavioral data.

Human infants behave in ways that appear notoriously present oriented. They seem concerned only with having their immediate needs met. By someone else. Right now. And once they can walk, that's when the trouble really begins. Reaching for hot frying pans, leaning over running bathtubs, charging recklessly onto the road. Toddlers frequently fail to foresee potential dangers.

Adults, by contrast, tend to instantly see when a situation is dangerous—that a candle is far too close to the curtains, or that running with those open scissors may end in a trip to the hospital. Parents spend considerable time pointing out obvious dangers and may despair at the need to remind kids of the same risk again and again. In our laboratory, we recently began to examine when young children start to recognize potential mishaps and act to reduce the chances of accidents. We put the children in situations with obvious risk potential, akin to inviting them to play basketball next to a vase or to do some painting atop a new carpet. Then we gave them ample

Figure 3.2. What could possibly go wrong?

opportunity to avert the potential accidents, by moving the vase out of the way before the first free throw, for example, or covering the carpet before dipping the brush into the paint. We also asked the children what could happen to others in similarly risky situations, such as what might go wrong for the girl in a short video eating spaghetti Bolognese in a clean white dress, or for the little boy leaving marbles lying around in a doorway. Five-year-olds were very poor at identifying hazards, and competence only gradually improved with age. While foresight is a quintessential human trait, it does not come on board wholesale, even for the most essential functions like dealing with risks.

Complicating matters, when children do act in ways that decrease risks, this need not necessarily reflect that they have thought about the future. A child may have been taught to put on an apron before painting and continue to do so out of habit without considering what could happen otherwise. So experimenters must devise careful procedures to ascertain when children are really acting with the future in mind.[23]

Aiming to rule out alternative explanations, we presented young children with a problem in one room, such as a puzzle box that, with the right kind of tool, could be opened to reveal a reward. Next, we took the children to another room and distracted them with other games for a little while. We then told them that they were about to return to the first room and that they could take with them one object from among a selection. Three-year-old children selected objects randomly (and sometimes, of course, picked the useful one by chance rather than by design). But, by age four, kids selected the tool that would help them solve the upcoming problem more often than expected by chance alone.[24] These choices demonstrate that they can remember a past episode sufficiently to recognize a required solution—and to secure it for a return to the problem. Such results have established that by age four, many children do have the basics of an inner time machine, enabling them to prepare for what they have reason to expect lies ahead.[25]

And when they expect they will fail at a future task, children from about this age onwards can use workarounds. In one of our studies, we introduced children to twenty-five upside-down cups, placed stickers under some of the cups, and, after a short delay, asked the kids to find the stickers. Unsurprisingly, they often lost track of where the stickers were hidden. Then we gave the children a chance to mark the target cups with tokens during the wait, or we gave them a pen, to see if they could come up with their own solution. Only from around eight years of age did children use the pen to mark the target cups (like drawing an X on a treasure map to mark the spot).[26] But even four-year-olds used the tokens more frequently when there were five target cups compared to when there was only one, presumably because they realized their chances of losing track of the stickers were higher when there were lots of targets.[27]

Four-year-old children also begin to show signs they appreciate *when* events will occur. They demonstrate some competence at placing daily happenings in appropriate order on a basic spatial timeline, such as putting breakfast before lunch and dinner, and they can distinguish daily affairs such as brushing their teeth from more remote future events, such as their next birthday.[28] However, at this age they tend to still find it difficult

to distinguish between annual (next Christmas) and more far-flung occurrences (when they might get married) on the timeline.

The developmental psychologist Cristina Atance and her colleagues
have documented many facets of how young children become increasingly
capable of taking potential future events into account. One of her early
studies showed that by the end of the preschool years, children could identify useful objects to take to a pictured place—such as bringing a winter
coat to a snowy environment—whereas younger children often selected
objects that would not be useful. Similarly, older preschoolers begin to
appreciate that they will like different things as grown-ups, such as preferring cooking shows over cartoons. Yet, in another study, many four- to
six-year-olds put candy in a room to eat tomorrow, even when they had
been told the room would be emptied overnight by a cleaner.[29] This suggests that though young children can act with the future in mind, it takes
time before they can reason about intervening events that might call for a
change of plans, such as the threat of a cleaner removing their stash.

As the various components of their virtual time machines increasingly
mature, older children and teenagers become more and more capable of
thinking ahead, and of controlling the future in line with their desires.[30]
The universal acquisition of these powers suggests that this is part of what
it means to be human. And yet, there are also meaningful differences, even
very early on in life, in how people steer their mental time machines, and
this appears to have dramatic consequences for the trajectory of their lives.

———

The social psychologist Philip Zimbardo has championed the idea that people can be classified as being primarily oriented to the past, the present, or
the future. Based on a questionnaire Zimbardo developed with his colleague
John Boyd, the past and present "time perspectives" can also be subdivided
into two more categories.[31] People who focus on the past may do so with a
positive sentimentality (agreeing more with statements like *Happy memories
of good times spring readily to mind*) or with a negative view (*I often think of
what I should have done differently in my life*). Those who focus on the present
may do so in a hedonistic manner (*It is important to put excitement into my*

life) or in a fatalistic way (*Fate determines much in my life*). Some scholars suggest a focus on the future may also involve subcategories of more negative outlooks (*Usually I do not know how I will be able to fulfill my goals in my life*) or more positive outlooks (*I complete projects on time by making steady progress*).[32]

A large body of research has documented associations between people's responses to such questions and their behavior, ranging from smoking and dieting to career choices and life satisfaction. Perhaps unsurprisingly, negative past orientation has been linked to depression, and present hedonism to taking drugs and getting speeding fines. A future time perspective predicts characteristics such as high educational achievements, conscientiousness, and choosing to delay gratification.[33]

Apparent variations in time perspectives emerge quite early in development. Consider the classic marshmallow test, in which the psychologist Walter Mischel and his colleagues gave kids a simple choice: have one tasty treat now or wait a while and get two instead. The experimenter would then leave the room, and the children would commence their struggle with temptation. Remarkably, children who managed to wait longer for the second marshmallow ended up doing better at school and on diverse measures of academic, personal, and social success in later life.[34] One reason for this might be that the same kind of self-control required to resist a snack applies in various other domains too. Life is full of trade-offs between rewards we can get now and those we must wait for or work towards. Being able to resist temptation and pursue longer-term goals instead would seem to be advantageous in many ways.

The idea that the marshmallow test measures something critical about children's later successes spawned somewhat of a cottage industry in psychology: for decades, researchers have set about documenting the virtues of delaying gratification and the perils of impulsivity. Through this lens, preferring to indulge in the moment represents a failure: a shortsighted psychological hitch to be smoothed out with cognitive interventions, reprimands, and self-help books.

More recently, however, the developmental psychologist Celeste Kidd and her colleagues added a twist to Mischel's paradigm, with results that challenge this conventional story. Before the marshmallow test commenced, an experimenter invited preschoolers to complete a fun art project. The experimenter then admitted that, unfortunately, the only art supplies available were a bunch of blunt crayons and some tiny, boring stickers. But, *aha*, just wait a minute, and the experimenter would come back with some brand new, fancy materials. There was a catch, of course: in only one condition did the experimenter return with what was promised. In another condition the experimenter returned empty handed, offering nothing but an apology. When it came to waiting for the extra marshmallow, children were no dupes. With a reliable experimenter, they waited patiently for an average of twelve minutes. But those with the unreliable experimenter didn't try their luck: they waited a meager three minutes on average before munching the marshmallow in front of them. And fair enough too. After all, untrustworthy experimenters may again fail to live up to their pledge.[35]

In general, there is always a risk that another person might eat your promised marshmallow until it is securely inside your mouth. Uncertainty abounds: owed money, budding romances, career opportunities, and other anticipated goods are obscured by the fog of the future. Take it from the ice-cream shop sign: "Life is uncertain, eat dessert first." Plus, aside from the risk that your cherished opportunity might not actually come about, there's also a chance you won't even be around to cherish it. In one study from our group, teaming up with the behavioral scientist Gillian Pepper, lower life expectancy in a given country predicted the tendency to select a smaller, immediate reward rather than a bigger but delayed one.[36]

Blue Eyes, a young offender from the streets of Atlanta, mused in an interview with criminologists: "I say f*** tomorrow. It's all about today. Might not be a tomorrow. Might get shot. Might get hit by a bus. So get it now. Now, now, now."[37] If there is no tomorrow, then today matters a lot more. In one study, economists invited followers of the evangelical Christian radio talk show host Harold Camping to choose between $5 right now or larger amounts of money to be delivered in a few weeks' time. Camping and his followers had long professed that "the Rapture" was due to occur on May 21, 2011—a fiery end to all days. To survey the devotees, the economists waited outside one of Camping's bible classes a couple of weeks before this purported earthly termination point. All but one of the twenty-three acolytes stated unequivocally that they would prefer $5 today to any amount up to even $500 payable one month later.*[38] It would be an error to call such decisions *impulsive* or *shortsighted* just because they focus on immediate rewards (even if they are based on a false expectation).

Yes, sometimes delaying gratification is the better move—but sometimes it is not. Perhaps more important than the ability to delay gratification per se is the ability to look into the future and determine one's best route forward: to use foresight to flexibly adjust decisions in the present. Given that the world did not end on May 21, Camping's followers better have updated their approach: presumably $500 in a month would have been quite enticing come May 22. The intuitive view that future orientation is better than a focus on the present, that delaying gratification always trumps impulsivity, must be tempered by questions about the right fit to the circumstances. The key is to learn when best to pursue long-term goals and when to focus on the present.[39]

Given that humans can so readily change their views and behaviors based on their current circumstances, it follows that classifying people into fixed categories of enduring attitudes towards time may not be entirely justified. Some people may be more focused on the present and others may spend more time pondering the past or planning for the future—on average—but that does not mean they cannot change their perspective when the situation calls for it.[40] Alongside other features of our peculiar psychology, foresight enables a huge amount of plasticity, which in turn underpins a staggering array of individual differences in human behavior from the earliest stages of development onwards. Next, we will see just how the development of foresight opens the door for people to pursue a diversity of different skills, knowledge, and abilities.

———

* The one follower who stated he would consider giving up $5 now to win a large sum next month also espoused "less than full beliefs" in the prophecy.

What we hope ever to do with ease, we must
learn first to do with diligence.

—Samuel Johnson (1791)

"He's what Tiger Woods is to golf and what Michael Jordan is to basketball," said Bill Orrell about his sixteen-year-old son, William, in 2015. "It takes a lot of focus and determination to do what he does." While growing up in North Carolina—just like Jordan—William spent at least an hour every day practicing his craft. In time, his achievements would exceed even his own wildest expectations. Between 2013 and 2018, he broke thirty-six division and world records, won three gold medals at the Junior Olympic Games, and appeared in a TV commercial alongside LeBron James. His skill? Speed stacking, a competitive sport with one primary objective: stack and unstack a series of plastic cups in a predetermined sequence as quickly as possible.[41]

Humans from across the globe become remarkably skilled in an extraordinarily wide range of domains. The Guinness Book of World Records is littered with obscure achievements that must have required thousands if not tens of thousands of hours of dedication. You can make up your own hitherto unclaimed specialty if you like. While there is already someone who holds the record for running the hundred meters on all fours (Kenichi Ito scampered this distance in just 15.71 seconds), and for the farthest bull's-eye shot by bow and arrow with their feet (Brittany Walsh hit the target from over thirteen yards away), there might still be opportunities if you started practicing running on two hands and one foot, or using your feet to, say, hurl a discus afar.[42]

Acquiring outstanding expertise often tends to be effortful and costly—at the very least because it means we cannot spend that time on other rewarding activities.*[43] K. Anders Ericsson, an expert on expertise, long argued that "deliberate practice" is the single most important factor in determining

---

* Ito, a businessman from Tokyo, claimed to have spent nine years studying the locomotion of quadrupedal animals—modeling his running style on the nimble patas monkey—before setting the hundred meters world record in 2015. When he worked as a janitor, Ito practiced this monkey running technique by mopping floors on all fours.

the acquisition of proficiency. He found that highly skilled young violinists had a higher number of practice hours under their belt than lesser-skilled violinists. Ericsson estimated that they would have practiced just over 10,000 hours on average by the time they were only twenty years old. Given that an hour every single day only gets you to 3,650 hours in ten years, it is clear that much of their youth was dedicated to learning this skill. Based on this research, the journalist Malcolm Gladwell suggested that a "10,000-Hour Rule" was the key to mastering a wide range of skills.[44]

This "rule" has become embedded in popular culture, no doubt partly due to the attractive implication that anyone can become fabulously skilled at something if they just put in a magical round number of hours. We often encourage our children with the romantic sentiment that they can readily achieve whatever they put their mind to. It is nice to be told *you can do it*. The flip side of this coin, however, is that it appears to place the responsibility for your shortcomings squarely on your own shoulders. If you do not excel, it must be because you failed to practice enough. It is not so nice to be told *you were just too lazy*. Reading this may bring to mind your own long-lost aspirations. Many of us have musical instruments and other gear in the house that attract dust and thoughts of inadequacy.

Was it just laziness that made you fail? It may come as a relief to learn that popular claims about the primary role of hours spent practicing are not quite right. Many people do not reach elite levels of performance in spite of putting in the requisite work, and conversely, equally excellent performers often differ in how much they practiced. A replication study of Ericsson's original work could only find a much smaller effect of the sheer number of practice hours on skill level.[45] And a review of studies on expertise in chess and music found that only about one-third of the differences between accomplished people could be explained by variations in how much they had practiced. Similarly, a study on performance in sports found that only about 18 percent of the difference between top athletes was due to differences in practice hours, and this effect was even smaller among the very top performers.[46]

So perhaps it does require at least a certain dose of innate talent to become a professional athlete or virtuoso musician after all. And, surely, it is a mark of intelligence to know what to pursue and what not to pursue given one's innate attributes. If you are five-foot-nothing tall, then you may want to put down that basketball and instead pursue a sport where your baseline attributes give you an advantage rather than a disadvantage. Though, of course, everyone likes inspiring stories of an underdog who makes it against all odds. Muggsy Bogues stood little more than five feet tall, but he played fantastic basketball in the NBA for fifteen years.

But let's look beyond the very narrow domain of elite performance and world records. Once we broaden our view, it becomes apparent that foresight unlocks for everyone the potential to learn ordinary skills to ordinary levels. Certainly, the primary difference between being able versus unable to play an instrument, a sport, or a game—let alone to type, drive a car, or dance the waltz—remains practice. The human ability to acquire new capacities, however banal they may seem, proves vital.

———

When someone's heart stops beating, it is possible to keep blood circulating through the body by putting pressure repeatedly on the sternum. Current recommendations state an untrained bystander should give about one hundred chest compressions per minute. Kneel next to the person's neck and shoulders, place the heel of your hand over the center of the chest, put your other hand on top, and keep your elbows straight with your shoulders directly above your hands. Using your upper body weight, push straight down on the chest until it compresses two to two and a half inches. The chance of surviving a heart attack is much greater if someone with this knowledge is nearby—better still if that person has been trained in CPR and practiced it. In that case, every thirty compressions would be interrupted by rescue breathing to add oxygen to the blood, and the victim would stand a better chance.[47] It is never too late to learn CPR; someone's life may depend on it.

By deciding what to learn, humans actively create the capacities their future selves will be equipped with. Although many basic skills, such as how to cook an egg or how to read and write, may be acquired by most people, particular repertoires of capacities vary dramatically between individuals even within a given culture. When you meet people for the first time, it is impossible to know the nature of their skill set. Are they musical, tech savvy, good at football? Can they whip up a mean salad dressing, change a tire, or perform skilled first aid? Although there are certainly major differences in talent and predisposition, the primary reason for this astonishingly unpredictable diversity is the time and effort that people spend honing their chosen skills and knowledge. Human beings are polytalented skill shifters, able to alter future capacities to suit their anticipated needs, creating a kaleidoscope of capability.[48]

Our research team has recently been examining when children begin to recognize that they can shape their future skills. In one of our studies, we introduced three-to-five-year-olds to a number of simple motor skill games, such as a task requiring them to guide a wire around loops of various shapes. An experimenter then told the children that they would be rewarded if they performed well on only *one* of these tasks—the triangle-shaped loop, for example. When the kids later had a chance to practice, only the older children tended to repeat the specific task they would be tested on in the future.[49]

We also told children stories about two characters who were going to play tennis, one of whom had played regularly and the other who had played only once before. By age five, children recognized that the frequent tennis player was probably going to win—and could explain why. Similarly, when asked what they would do if they wanted to get better at something, they tended to mention practice and rehearsal. The same children who spontaneously mentioned the benefits of practice were also the ones most likely to have picked out the practicing character in the story task, and to have practiced themselves in the motor skill game. By this age children demonstrate an explicit understanding of the link between deliberate practice and skill acquisition across a range of contexts.[50]

Practice, of course, tends to involve more than just repetition. If you want to be able to juggle, you cannot expect to just pick up three balls, toss them around for some time, and join the circus. Learning a motor skill tends to involve repeating small actions (such as tossing one ball from one hand to the other) until these become routine, and then adding another action (tossing one ball with one hand and then tossing another ball with the other hand as the first ball reaches the top of its arc). Eventually, a more complex action sequence can be learned (juggling three balls), until that too becomes automatic.[51] In turn, combining and recombining smaller action sequences enables people to create something new. A circus juggler can create novel routines, similar to how a practiced guitarist can readily combine a finite number of basic chords and strumming patterns into a potentially infinite number of different songs.

Deliberate learning is of course not only directed at enhancing what we can do. Humans also deliberately enhance what we know. The psychologist Irving Biederman has called humans "infovores" because of our pervasive hunger for new information.[52] We are curiously curious animals, almost voracious in our appetite for news (indeed, the German word for curious, *neugierig*, literally means "news greedy"). In another experiment from our team, children were presented with one red set and one blue set of cards decorated with cartoon animal characters. On the other side of the red cards was, say, a picture of that animal's favorite food, and on the other side of the blue cards was its favorite toy. Children were then told that they could win stickers by later recalling the animals' favorite toys (but not foods) and were given one minute to look at what was on the other side of whatever cards they wished to turn over. Most of the four- and five-year-olds turned over cards from both sets, but six- and seven-year-olds spent nearly all of their study time examining what was on the other side of the target cards rather than of the distractors. These emerging infovores seemed to know what information they needed to find and remember.[53]

Of course, children are usually not left to their own devices in shaping their knowledge and skills. As we saw, they are taught the facts and skills adults anticipate they will need to learn, even when the children themselves have other priorities. This may include facts about how to improve skills: *practice makes perfect*. A teacher can demonstrate how it's done, identify errors, and guide the way forward. In skill learning, an initial focus on getting a pupil to copy the actions of the teacher tends to gradually give way to a focus on producing the desired outcome. And so *watch my hands* makes way for *play an A minor and then an E major*. In many cases, children persist with studying on their own accord. It may be no coincidence that formal schooling tends to start around the age of six, given that children may only then be establishing the cognitive foundations we have seen are essential for deliberate, farsighted practice and information seeking. Ultimately, deliberate practice is essentially teaching oneself, and so in a sense teaching and practice are really two sides of the same coin.[54]

While studying and practicing can at times be fun and rewarding in the here and now, deliberate learning is often difficult and frustrating—even downright embarrassing. Beginner guitar players may suffer through months of busted strings and broken skin before they can play even the simplest songs. Somehow, the motivation to persist must be stronger than the disincentives. And it must also be strong enough to compete with more immediately gratifying urges—such as putting on a record and listening to Jimi Hendrix do it better than you ever could.

To motivate us to prioritize future-directed learning, our mental time travels have to be emotionally charged. It pays to feel good when imagining shredding that guitar solo, just like it pays to feel good when imagining attracting the partner of your dreams (perhaps helped by an outstanding performance). Conversely, it pays to feel bad to imagine flopping out in such endeavors. The longing and the envisioned humiliations drive us on.[55] Indeed, research suggests that we regularly exaggerate the emotional importance of future events: we systematically overrate the happiness we expect to get from achievements and the negative consequences of future failures (you can look forward to more on this in the next chapter).

Aside from driving us to pursue longer-term goals, our imagination can itself do the job of helping us reach them. One of the most remarkable features of the human mind is that we can practice purely in our heads, rather than, well, in practice. Instead of repeatedly engaging in the relevant action sequences, merely imagining performing those same actions can be effective. For instance, actually practicing playing a piece of music on the piano and imagining playing the piece in one's mind both result in improved performance compared to not doing either.[56] And as many sportspeople will tell you, mental rehearsal of certain skills is far less likely to hurt you in the process—just think of aerial ski jumping.

Contrary to some outrageous popular claims, however, simply envisioning being successful—say, picturing yourself on the podium popping champagne—is unfortunately not going to improve the necessary skills. In one study, college students were asked to either imagine getting good grades on their exams or to imagine studying for the exam. Those who focused on the process significantly improved their grades compared to those focused on the outcome.[57] Research by the psychologist Gabrielle Oettingen suggests that it can even be counterproductive to fantasize about positive futures.[58] Instead, it may be better to picture what you are going to do if things go wrong or if you encounter obstacles.

Repeated testing is one very effective method to help you remember information when needed—much to the groans of students everywhere. Yet knowledge consolidation can also take many other forms, such as with mnemonics that group large amounts of information into smaller, memorable chunks that can later be unpacked. Anticipating that we might forget what to do when the critical time comes, we put these approaches into practice. In a medical emergency, for example, you should first consult DR ABC: check for Danger, ask the patient for a Response, make sure the Airway is clear, and then look for problems with Breathing and Circulation. In the absence of breathing or circulation, you should begin CPR.[59] If you have already forgotten how frequently you should compress the chest of the patient—we told you at the start of this section—then here is a way to help yourself remember. Do it to the beat of "Stayin' Alive" by the Bee Gees.[60] It comes with a very apt mnemonic—*Ha, Ha, Ha, Ha.* Or, if you prefer something less optimistic, use Queen's "Another One Bites the Dust."

———

We all benefit from the knowledge, skills, and products generated by others, such as experts in medicine, craft, technology, science, and education. We trust our pilot knows how to land the plane. We can ask a carpenter to make our chairs, a baker to make our bread, and teachers to help us learn. As a species, humans have acquired a far wider variety of expertise than any other animal. This has empowered us to adjust to countless environments and circumstances. Modern humans live from the tropics to the arctic, on islands and mountains—and, as we saw in the last chapter, depend critically on the culturally maintained expertise that each new generation must master.

Other animals do practice, mind you. For instance, young mammals often play in ways that resemble real fight-or-flight patterns—allowing them to incidentally fine-tune their skills. It is easy to see how natural selection would favor individuals with such predilections when they are later confronted with the serious struggle for existence.[61] Humans also learn many skills incidentally, as side effects of engaging in other activities, such as when children improve their hand-eye coordination by playing throwing games. However, there is nothing to suggest that other animals spontaneously rehearse and refine their movements to deliberately improve particular skills. Consequently, individual animals of the same species do not show the same unpredictable diversity of capacities that humans readily demonstrate. A duck is a duck is a duck. Humans can, of course, train animals and get them to practice through standard conditioning techniques. Through reinforcement (and selective breeding), a dog may be shaped into a herder, guard, hunter, retriever, or drug detector. But it is the human, not the dog, who looks ahead and deliberately sets out to forge those skills with future expertise in mind.[62]

Foresight confronts us with choices about our own future selves. These choices are not limited to deciding which skills and knowledge to acquire. We are concerned about many other aspects of our future selves, including how others will see us—even just literally. As well as being skill shifters, we are also shape-shifters—like the polymorphs of science fiction—mutable in our appearance. Many people prepare for the day ahead by selecting what clothes to wear, and with a look in the mirror, they apply creams, fragrances, and makeup, feeding a massive global cosmetics industry that is valued at over $500 billion.[63] The world over, people also choose to shape their bodies in more lasting ways, as they wear braces to straighten their teeth; stretch, pierce, and tattoo their skin; cut, style, or pluck their hair; and go on diets or bulk up their muscles with exercises and supplements. As cosmetic surgery has become more sophisticated and affordable, more and more people have breasts enlarged, noses shrunk, or ears pinned back. People are often driven to such interventions by the desire to become more comfortable in their own skin, to express themselves, to become more attractive to others, or to reverse the inevitable signs of aging—at least for a few more years.

Other modifications are aimed not at how we appear to the world but rather at how the world appears to us. We deliberately enhance our senses and compensate for our fallible bodies. Contact lenses and glasses are installations that let us see farther and more clearly, and cochlear implants can bypass a damaged auditory system to deliver electric signals straight to the critical nerve. In the future, we will likely be able to modify ourselves even more, with biotechnology, nanotechnology, and other "-ologies." Just consider a small device called NorthSense, developed by the British company CyborgNest, that can be permanently attached to the upper chest. It vibrates slightly when facing magnetic north. This allows wearers to feel where they are in space relative to the electromagnetic field of our planet, perhaps not unlike how migrating birds do.

At the heart of this obsession with changing appearances, senses, and capacities is our ability to envisage our personal futures. We see our future selves as at once continuous with who we are today, but also potentially different.

Finding out what we want to do with our lives—and the type of person we want to become—can be one of our biggest conundrums. Psychologist Abraham Maslow famously claimed that humans have a deep need for *self-actualization*, or for striving towards becoming the best possible person one can be. This concept has faced criticism from other academics, who claim that it is difficult to research with scientific rigor.[64] After all, the idea of self-actualization implies that there is an optimal way for an individual to live, and to measure it would require impossible knowledge of all the outcomes of all the different courses of action that one could have chosen to pursue but did not (maybe you would have been a fabulous circus juggler after all). Still, the idea has seeped into popular culture, probably because the general notion of self-improvement—within the boundaries of one's innate potential—is so universally relatable. We have all had to face the question of what we want to be when we grow up; some of us are still wondering.

———

We have seen how children only gradually assemble their inner time machines. While practically everyone acquires the basic component capacities by the end of the preschool years, foresight continues to mature over childhood and beyond.[65] With the gradual development of mental time travel, children discover that hopes and expectations can be disappointed, that plots and plans may not always come to fruition. So they learn to make contingency plans and employ workarounds. But uncertainty about the future can still cause great anguish. In the modern world, we may even feel paralyzed by the great variety of possibilities and choices. Famously, Jean-Paul Sartre said that humans are "condemned to be free."[66]

People often end up looking back at past choices with regret. In studies from our laboratory led by Shalini Gautam, we found that young children experienced regret when they picked one of two boxes and won a prize but were then shown that choosing the other box would have led to a larger prize. The children only felt sad if they could have picked the other box, though, and not just when we showed them what was inside the other box without having given them a choice. Just like adults, the children felt regret because they imagined a counterfactual past decision and compared it to their actual past decision.[67] Regrets, though often lamented, are tremendously useful for a developing time traveler. They motivate us to learn from the past and drive us to make different decisions in the future.[68]

The development of our virtual time machines gives us great powers but also presents us with endless new challenges. Children must learn to make decisions about what to pursue and what to prioritize. Most of us manage to acquire competences in many ordinary domains and often excel at least in some areas, however peculiar. But given that becoming a master in certain skills can take considerable dedication, people cannot hope to achieve elite levels in every domain (even if there are some polymaths with very impressive lists of abilities). There is only so much time in a day.

We tend not to leave it entirely to children to shape their own future. At first, children's decision-making tends to be limited to minor and immediate choices as parents take care of the essentials and longer-term strategies. Societies benefit from guiding children's learning so that they all acquire the basic competences they can draw on for the rest of their lives. As was the case with innovations, it is the recognition of future utility that drives parents, institutions, and eventually children themselves, onwards. Our cultural heritage includes many carefully designed curricula that ensure key lessons are learned—and curricula themselves are continually refined to keep up with changing times.

Foresight and culture also guide the learning of specialist expertise. If you want to become an electrician or a nurse, you will have to do specific courses and sit specific exams, thereby ensuring that you will be prepared to deal with the challenges other experts in these fields encountered. In turn, we can all depend—more or less—on the diverse expertise acquired by those around us. Today, thanks to specialist medical know-how and the reliable transfer of knowledge and skills, we can expect anesthesia during surgery and survive heart attacks where other animals would have invariably bitten the dust.

Most animal species appear to be either *specialists*, like koalas and dung beetles, adapted to very specific environmental opportunities, or *generalists*, like rats and pigeons, capable of coping with many diverse habitats. The skill-shifting characteristic of the human brain, by contrast, enables us to be both. We are *generalist specialists*, if you do not mind the apparent oxymoron.[69] Human children gradually acquire the generalist capacity of foresight, which in turn enables them to specialize in preparation for upcoming challenges. We can be trained—or train ourselves—to become diverse experts, even if we acquire mainly common skills to mediocre levels. Cultural evolution interacts with our individual foresight, and this has enabled us to create societies comprised of complementary skills and knowledge, with people that incessantly cooperate in ways that allow the group to benefit from its members' expertise. Humans adapt not only through the blind watchmaker of natural selection. We have evolved brains that allow us to see ahead and invent ourselves.

CHAPTER 4

UNDER THE HOOD

It has become abundantly clear that human behavior is active in character, that it is determined not only by past experience, but also by plans and designs formulating the future, and that the human brain is a remarkable apparatus which cannot only create these models of the future, but subordinate its behavior to them.

—Alexander Luria (1973)

A sixty-two-year-old man dies, unexpectedly, in his assisted-living home. His mother gives a team of scientists the go-ahead: his brain is removed, bathed in formaldehyde, scanned with a high-tech magnetic resonance imaging machine, cut up into slivers (some many times thinner than the width of a human hair), mounted on glass slides, stained with more chemicals, and photographed in ultra-high resolution.

The man's name was Kent Cochrane, and when he was thirty years old, he was in a motorcycle accident that severely damaged his brain. After the crash, Kent was unable to remember any event from his life whatsoever. Not his exciting youth spent driving homemade dune buggies, rocking out in a band, and staying up late playing cards in bars; not even the moment he heard about the death of his brother two years before his own accident.

When asked what he might do *tomorrow*, Kent would also draw a blank. Once, when the eminent psychologist Endel Tulving asked him for an analogy, Kent said: "It's like swimming in the middle of a lake. There's nothing there to hold you up."[1] Alongside his past, Kent had lost his grip on the future.

In March 2020, the results of the painstaking brain analysis confirmed what neuroscientists had long supposed. Kent had damage to various parts of the brain. His *hippocampi*, two structures housed deep in the temporal lobe of each hemisphere, had been almost completely destroyed.[2] Clive Wearing, whom we met in Chapter 1, also suffered hippocampal damage, though in his case the culprit was the herpes simplex virus that had found its way into his nervous system. By studying people like Kent Cochrane and Clive Wearing, as well as healthy brains in action, scientists have begun to chip away at the question of how a wet and squishy three-pound clump of cells builds a picture of tomorrow.[3]

The human brain is arguably the most complex machine in the known universe, even if we can run it on a sandwich and a cup of coffee. It contains billions of neurons—specialized cells that communicate via electrochemical signals—organized into a dizzying array of networks and circuits. The relationship between the brain and the mind has of course long persisted as one of the deepest of all philosophical problems.[*4] Nonetheless, as cases of amnesia have amply demonstrated, a wrench in the biological gearwork also changes the mind. Our abilities for memory and foresight, like everything else in cognition, appear to arise from the activity of matter: the cells of our nervous systems and their interaction with the rest of the body and the world around us.[5]

For much of the history of neuroscience and psychology, researchers found it useful to think of the brain as a primarily reactive organ. An object appears in front of us, its image is projected into the eyes or its sound into

---

* René Descartes's famous assertion of mind-body dualism—that the mind and body are two different substances—captures a common intuition. Many of us reach the conclusion there must be some deep divide between the stuff of thought and the stuff of matter. Regardless of how much evidence accumulates on the physical foundation of cognition, people tend to be "common sense dualists," as developmental psychologist Paul Bloom put it.

the ears, and these signals then get piped up through the brain, processed layer by layer, until a sensory perception results. Only then is this information used to marshal a response. But in recent years that model has been turned on its head.

There is now little doubt that brains are much more proactive than had been assumed, constantly generating predictions to deal with tasks regardless of their location in time—including, somewhat paradoxically, perceiving the present. Processing sensory input takes some small amount of time, let alone figuring out what to do about that input.[6] If you always had to wait for input to be fully processed after it entered through the senses, only then sending a command to the muscles for action, you would be permanently living a fraction of a second in the past. But evolution is not very tolerant of slowpokes.[7] A snake would have a mouthful of your ankle before you even noticed it raising its head.

To live in the present, our brain must continually forecast what's coming next. Even apparently simple tasks require some pretty sophisticated predictive processing. Catching a gently thrown tennis ball may feel effortless, but when a ball gets tossed in your direction, you need to act fast: predicting a precise trajectory so your hand can find its way to exactly the right place at exactly the right time.[8] As we will see in this chapter, there is a growing consensus in neuroscience that prediction is more than a quirk of cognition. Instead, it is the neural modus operandi—a core feature of what makes the nervous system tick.[9] Prediction is not only involved in perception and motor coordination but also manifests in our remarkable capacity to run simulations of tomorrow and beyond.

———

Imagine you are watching one of your favorite bands playing a concert. The dark venue is flashing with colorful lights, it's noisy, it's crowded, and you're stuck somewhere near the back. As heads bob back and forth before your eyes, you try to see what's happening onstage: *Which guitarist is playing that solo? Is that the regular drummer, or have they brought in a guest?* Each glimpse you catch of the stage gives you a little bit more information, and eventually you piece things together. You can figure out what's happening onstage, despite your limited visual access, because your brain is a good guesser—it builds a mental picture of what's going on around you.

You face a constant barrage of ambiguity, even when just engaged in routine activities like getting a cup of coffee. There is something going on *out there* in the world, but the brain, cradled snugly away in the silent darkness of the cranium, has no direct access to whatever that is. Instead, it must use clues in the form of data it sucks up with its sensory organs, such as your eyes, ears, and nose. The problem is that sensory input often has any number of different interpretations. If you look at a flat, circular object like a coffee coaster from the side on, your retina detects only a thin, straight line. With no context, a thin, straight line could be any number of different objects: a small book, a straight comb, the edge of a photo frame, a sideways smartphone. But, of course, your brain usually does have at least some context: you are in a café, you can smell the coffee, and there is a barista about to put this object on the table in front of you.

The pioneering nineteenth-century physiologist and physicist Hermann von Helmholtz was among the first to flesh out the idea that even the simplest act of perception requires us to make educated guesses about what is happening in the world around us.[10] Helmholtz saw perception as a process of inference because the brain doesn't just get the true state of the world channeled into it passively through its sensory systems. Instead, in much the same way as the group of blind men in the old Indian parable can eventually figure out the shape of an elephant by each touching a different part of its body, sensory systems use dribs and drabs of data to infer the presence of guitarists and drummers, coffee and coasters.

How does the brain make these best guesses about the causes of its incoming sense data? One intriguing possibility is that it works much like a particular kind of statistician. In Bayesian statistics, named after the Presbyterian minister Thomas Bayes, probability is framed in terms of how strongly one believes in the occurrence of some event. Let's say you are going to toss a coin one hundred times. You start off thinking that the coin is fair, so you firmly believe you will get about fifty heads and fifty tails. Yet you find that ninety times out of one hundred the coin comes up heads. The handy thing about Bayesian reasoning is that you can work your way

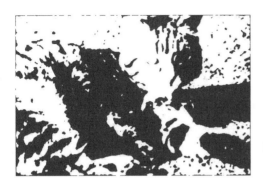

Figure 4.1. What do you see?

backward from the data to update your belief about the hidden causes of what's going on, and about what will happen next: the coin is probably rigged, and the following flip will likely end up on heads too. In a similar way, the brain appears to make predictions and then checks to see whether these predictions are borne out as new data roll in so it can update its subsequent predictions.[11]

Take a look at Figure 4.1.[12] Can you tell what it is? It probably looks like just a bunch of splotches.

Now turn the page and look at the photograph before coming back to this image.

Did anything change? If the shift in perception was successful, you will not be able to look at this splotchy version again without seeing the cat. The sensory input is precisely the same both times you look at the splotches, but your brain has received new evidence about what is being depicted. In turn, it has acquired a new interpretation of what is causing the splotches hitting the retina, which gets expressed as an updated visual experience.[13]

One of the most crucial mechanisms that allows this fine-tuning of perception is the selective use of error. To update its beliefs, the brain makes use of the difference between predictions and what then really happens. If you've ever popped an olive into your mouth thinking it was a grape, or waved across the room at a friend only to realize she was actually a stranger, then you know how it feels to make an error of sensory prediction. Very rapidly, you come to expect the next appetizer to be savory rather than sweet, the stranger to give you a puzzled glance rather than a warm smile. These are conspicuous examples, but the brain is constantly managing far more subtle

Figure 4.2. Does seeing it like this change the previous image?

errors of prediction and relentlessly updating hypotheses about the causes of its sensory input.

Several lines of evidence support the idea that the sensory system gives priority to unexpected information. For instance, the cognitive neuroscientists Marta Garrido, Jason Mattingley, and their colleagues have demonstrated that mismatches between expectation and observation, such as a rare note in a sequence of tones, produce distinct spikes in neural activity even when we are focused on other tasks.[14] Predictive processing in the brain works much in the same way as computers do to save on storage space and speed up transmission.[15] Computer algorithms compress data in a video file not by encoding information about every single pixel—that would be absurdly wasteful—but by encoding only information about those specific pixels which have changed from previous frames. So too, the brain focuses on processing those aspects of the input that were not predicted. Much more efficient.

Our perception, then, is the result of what we expect to see and what information actually comes streaming in. Once you know it's a cat, the blobs are forever feline. Such predictive coding has been documented across sensory domains. But the capacity to perceive can only be part of the story of how brains function. Brains must also guide action.

March 19, 2016, the Four Seasons Hotel in Seoul. More than one hundred million people worldwide are watching as Lee Sedol readies himself for the match of his life. He is about to play Go, the ancient Chinese board game long recognized as among the most complex in existence. Sedol is a master, a world champion and one of the best players of all time. But his opponent is not another human. It is a computer program called AlphaGo, a set of learning algorithms built by the company Google DeepMind.

In Go, two players take turns placing black and white stones onto intersections on a large grid board, and the winner is decided based on whose stones encircle the largest amount of territory. The simple rules interact to produce a stupendous number of possible board configurations, more, in fact, than the estimated number of atoms in the universe.[16] Trained on games between human experts, and eventually by playing against itself millions of times, AlphaGo's algorithms rely on reinforcement learning. Actions can bring positive or negative outcomes, rewards or punishments. To maximize future rewards, a reinforcement learner repeats in the future whatever actions previously led to the most positive outcomes in similar situations. AlphaGo learned how to play Go well enough to win against Sedol four games to one—the first time a computer beat a world champion. Sedol retired from Go in 2019, saying that he could no longer bring himself to play because such a program "cannot be defeated."[17] The strength of artificial intelligence systems like AlphaGo derives from the powerful ways they harness the logic of learning mechanisms that biological brains have been exploiting for millions of years.

In animal brains, reinforcement learning relies on activity in cells called dopamine neurons. Dopamine is well known as a "pleasure" chemical because of its association with reward-related activity. But the reality is a bit more nuanced. Instead of dopamine simply signaling reward, dopamine neurons fire the most when an animal gets more reward than it expected (and the least when it gets less reward than expected).[18] The reward-prediction error signals that dopamine neurons transmit to other parts of the brain, much like the sensory-prediction error signals we just encountered, therefore, enable an animal to fine-tune its behavior. If you predict you will get some small amount of pleasure when opening a new macadamia nut chocolate bar, only to find the flavor combination to be absolutely delicious, this will involve a rush of dopamine and an update to your expectations. Next time you will be more enthusiastic about the prospect, and you'll know to seek out macadamia chocolate instead of other snacks.

This kind of reward-prediction error mechanism enables animals to efficiently learn about their environments and calibrate their behaviors to achieve desirable ends. Of course, humans and other animals don't just predict the world and learn from it: we want the world to be a certain way. We have goals.[19]

While we have already seen that the tissues of the hippocampus appear to be critically involved in traveling mentally through time, they are also key structures for travel through space—often in the pursuit of goals. In 2014, John O'Keefe, May-Britt Moser, and Edvard Moser received the Nobel Prize in Physiology or Medicine for discovering special neurons in the hippocampus and surrounding areas that comprise something like a GPS in the brain.[20] Some of these neurons, *place cells*, fire only when the animal is in a specific location. Other neurons, the aptly named *head direction cells*, fire depending on which direction the animal's head is pointing. Yet others are *border cells* that appear to respond to the presence of an environmental boundary at a particular distance and direction from an animal, or *grid cells* that fire at certain intervals to create and maintain a navigational frame of reference. Together, the coordinated firing of these cells creates little circuits that allow animals to represent their position in space—a mental map that codes for locations, distances, and directions. Such maps, built from single cells firing and wiring together, are the reason why an animal can so readily achieve its goal of getting from A to B, even if it has never taken that particular route before.[21]

Thinking about space is a good place to start thinking about time. As a rat moves around a maze, the corresponding hippocampal place cells for its location can sometimes fire in quick, short cycles. In a single cycle, lasting only around one-tenth of a second, place cells activate in a sequence that traces the path of the rat, from just behind to slightly in front of its current location. What's more, the neuroscientists Adam Johnson and A. David Redish have documented that "forward sweep" preplay sequences can also extend farther ahead of where a rat is currently standing and even alternate back and forth between two apparent goals—such as between a piece of strawberry down one tunnel and a piece of banana down another—on two different sides of a fork in the road.[22]

In the 1930s, psychologist Edward Tolman had already noticed that, on reaching such a fork, rats would look back and forth between the two routes as if trying to decide which way to go. Tolman suspected that the rats were imagining their options, in a kind of mental trial and error.[23] Now neuroscientific work has linked this curious behavior to the forward sweeping activity in cells of the hippocampus. When rats look back and forth between route options, it is not just a sign of exploration: the behavior occurs most often when the rats already know the layout of a maze but don't know where the food is hidden that day. Then, if food is found in the same place multiple times, the rats' head movements and place-cell forward sweeps gradually change, mostly in tandem. As a rat does more and more laps of the maze and the food is found in the same place each time, the animal increasingly only looks ahead—and the place cells only sweep ahead—to the path it eventually chooses. And, once the rat knows exactly where to find the food, it stops looking both ways at the choice point, and there are barely any forward sweeps in the hippocampus at all.[24]

Many contemporaries of Tolman dismissed the idea of mental trial and error in rats. Most were staunch behaviorists who thought that scientific psychology should be concerned only with observable stimuli and responses, rather than with something as unverifiable as the inner workings of the mind. One scholar even quipped that "so far as the theory is concerned the rat is left buried in thought."[25] But so far as Redish and other neuroscientists today are concerned, this is right on the money. As we will learn in the next chapter, where we tackle the question of animal foresight head-on, there is more to this story than meets the eye. Is it feasible to draw conclusions about what's going on in the mind of an animal by looking at its brain cells?

There are good reasons why researchers can't usually do this kind of single-cell recording work in humans. To measure the electrical activity of brain cells, scientists open the skull and insert electrodes into the brain of a living animal. The animal is still able to walk around after this procedure, allowing neuroscientists to measure, in real time, neuronal firing patterns as it moves through space. These methods obviously pose ethical dilemmas and call for a careful weighing up of the costs and potential benefits of their use. In some rare circumstances it is possible to directly measure signals from individual neurons in the human brain, and this can be illuminating in ways that animal research cannot. For example, people with epilepsy sometimes have electrodes implanted directly into their brains to record activity in their cells while they remain awake and conscious so that doctors can figure out the origin of their seizures.[26] Such measurements have revealed that, like those of other animals, the human brain has a mapping system consisting of place cells and grid cells. We also share *time cells*, which fire at particular moments of an experience and so appear to track the flow of time.[27]

In an ingenious series of experiments, the neuroscientist Rodrigo Quian Quiroga and colleagues further demonstrated that the human medial temporal lobe (home of the hippocampus) contains cells which fire in response to very specific concepts, such as "the Eiffel Tower" or "Jennifer Aniston."[28] An Arnold Schwarzenegger concept cell would fire whether you were hearing the words *Arnold Schwarzenegger*, seeing a picture of the former governor of California wearing a suit, or seeing a shot of the actor in *The Terminator*, kitted out with a leather jacket and sunglasses. The same cell would even fire when remembering a picture of Arnie's face, something we know because scientists have been able to directly ask participants to remember what they had seen previously, and then observe which individual neurons are active during both the original encoding and the subsequent memory.

The key thing about these concept cells is that they enable the encoding of abstractions: the meaning that holds disparate perceptual inputs together (we know it is the same Schwarzenegger regardless of whether he's hitting the gym, giving a press briefing, or wielding a shotgun). We can think about concepts in the abstract without the clutter of the specific memories. The decontextualized neuronal coding may be critical for imagining the future because it provides elements we can combine in new contexts—if he could run, what kind of president would Arnie be?

Humans benefit enormously from creating internal models that not only capture the causes of incoming sensory signals but also map out the causal relationships between possible behaviors and their outcomes. The human brain is an industrious cartographer: making maps of external reality, maps of conceptual relationships, and maps of the future.

The cognitive neuroscientists Daniel Schacter, Donna Rose Addis, and their collaborators have used neuroimaging techniques to examine brain activity when participants are directly asked to remember the past or imagine the future. One way to envision the future is to simply recast something from memory—imagining going to the supermarket tomorrow by picturing what happened last time. But as we have seen, human foresight lets us do much more than that. We can think about events we have never experienced before by combining elements from memory. The researchers first provided participants with a list of memories they had reported earlier—for instance, sipping a beer while watching your favorite band at a bar last year, going shopping for a bicycle with your friend Talia last month, and playing football with your team last week. Then the experimenters shuffled the details together and asked the participants to imagine future events involving random assortments, such as *Talia, bar,* and *football.* In the scanner, participants therefore imagined potential future events that they had never experienced, like going with Talia to watch football at a bar next month. It turns out that many of the same brain regions were active in this condition as when participants remembered actual past episodes.[29] Our mental time travels into past and future depend on much of the same neurocognitive machinery.[30] This is why people who have damage to that machinery, like Clive Wearing and Kent Cochrane, struggle to travel in both directions.

To confirm that participants are actually remembering or imagining as they are instructed in neuroimaging studies, experimenters can only rely on the participants' verbal reports, which are hard to validate.[31] Nonetheless, this approach has produced a number of consistent findings, including that, as people get older, their mental time machines produce fewer episodic details (the *whats, wheres,* and *whens*) of scenarios both past and future. Cognitive neuroscientist Muireann Irish has also documented, through her work with people with Alzheimer's disease and other forms of dementia, that brain atrophy associated with memory deficits is likewise associated with profound impairments in imagining the future.[32]

Ultimately, to do anything useful, brain activities must guide behavior. In various studies from our team, led by Amanda Lyons, Julie Henry, and Peter Rendell, we found that people with known neurological impairments to memory regions also show behavioral impairments in solving future problems. In a "virtual week" task, participants play a computer game where they encounter various problems and must take actions to solve them. Some of the problems cannot be solved right away, but participants are later presented with an opportunity to secure a solution for a future return to the problem. Say you have nothing left to feed your cat, and later on in the virtual day you go shopping and come across the pet food aisle. We found that older adults, people with substance use problems, individuals with schizophrenia, and people who had strokes all showed parallel deficits in memory for past problems and in securing the right items, such as the cat food, to solve those problems in the future.[33] These studies illustrate that foresight is not only affected in severe cases of amnesia; even less extreme declines in memory capacities are associated with impairments in imagining the future and acting upon it.[34]

To project yourself backward or forward through time, you of course need some kind of self to project. Even toddlers appear to have a basic awareness of who they are. They will use mirrors to examine parts of their own body they otherwise cannot see, and will remove a surreptitiously placed sticker from their hair.[35] Likewise, the great apes—chimpanzees, gorillas, and orangutans—have repeatedly passed these kinds of mirror self-recognition tasks. The species most closely related to us on the tree of life therefore share at least an ability to form an expectation of what they look like in the present, suggesting that whatever aspects of our brains are critical for visual self-recognition are shared with them.[*36] Of course, self-awareness in humans involves much more than recognizing our image in a mirror.[37] We have a unique collection of memories, plans, and ambitions, and—as we saw in the last chapter—an ability to shape our future selves deliberately. We can be aware of our likes and dislikes, our attitudes and personality, our strengths

---

* The ability may not be shared with our next closest relatives, however. In one study with Emma Collier-Baker, we sneakily smeared icing sugar on the legs of small apes (gibbons and siamangs). Naturally, they devoured it as soon as they discovered it. But when we instead put the icing sugar on their foreheads, meaning they could see it only in a mirror, the apes did not use their reflection to get to the treat. They tended to peer over or reach behind the mirror, as if searching for "the other ape." The fact that small apes consistently fail the mirror self-recognition task suggests that whatever brain mechanisms are responsible for visual self-recognition are shared by humans and great apes but not by our next closest relatives.

and weaknesses, and we can reflect on how all these things have changed throughout our lives and might change in the future.[38]

Psychologist Endel Tulving once said about Kent Cochrane that "in many ways, he may be happier than many others." The reason? "He does not worry about the future. . . . He does not worry about death."[39] With brains equipped for foresight, we find it difficult to look away from the fact that, at some point, our mortality will catch up to us and our selfhood will come to an end. Reactions to this realization take radically different flavors: denial, you-only-live-once hedonism, a search for divine salvation, stoic acceptance.[40] The realization of our mortality is a characteristically human phenomenon, a wellspring of human activity. It stems, ultimately, from our ability to wade virtually across time and space and construct a sweeping narrative—our own personal life story.

———

But let's meditate on the present for a moment. Meditation, as it is often practiced, involves sustained attention towards a simple phenomenon in the here and now, like the breath. According to the psychologists Ruben Laukkonen and Heleen Slagter, meditative practices work by gradually turning down our otherwise ceaseless predictive processes. Some highly trained experts claim to even experience a loss of their sense of self when entering into deep states of meditation.[41] Many beginners, however, are astonished to notice how often the mind flits away from the present moment. You may find it difficult to simply focus on your breath for even a few minutes without drifting into considerations about what's for dinner or whether your gym membership might be expiring soon. We are incessant mental time travelers.

Even when participants are simply asked to rest in a neuroimaging study, many of the same brain regions are as active as when they are asked to remember the past or imagine the future. Mental time travel is so fundamental that we tend to default to it when there is nothing else to focus on. Indeed, the interconnected brain regions that are most active at rest are known as the *default mode network*.[42]

Defaulting to daydreams can have some downsides. Mind wandering in a lecture, in the middle of reading, or while driving can lead to outcomes from the mildly annoying (having to reread the part of this book you just zoned out on) to the catastrophic (missing a stop sign). In one study, when experimenters asked people in an emergency room about their recent traffic accidents, highly distracting mind wandering was among the strongest factors associated with responsibility for the incident, alongside inebriation or sleep deprivation. On the other hand, as cognitive psychologist Michael Corballis highlighted in his book *The Wandering Mind*, allowing ourselves to explore away from the confines of the present can also be a potent source of inspiration—a form of "creative incubation." People may put down some problem, only to spontaneously come to a better solution later—perhaps in the shower or while on a walk. In laboratory studies, increased mind wandering has been shown to benefit performance in many different problem-solving tasks, such as planning how to make friends in a new neighborhood.[43]

Clearly, mind wandering has both costs and benefits depending on the circumstances. But at least we can catch ourselves in the act and attempt to control the stream of our thoughts. In one study led by the psychologist Paul Seli, participants were given a simple, boring task: they watched as the hand of a clock would take twenty seconds to make a full rotation, and just had to press the keyboard space bar every time the hand reached the upright position to win some cash. Easy enough—except that to win the money, participants had to press the space bar within half a second of the hand going vertical. Throughout the task, the experimenters inserted "thought probes" at various points. The clock would pause, and a single question would appear on the screen asking whether the participant was thinking about anything unrelated to the task. Minds weren't wandering at random. Instead, there was a fine-tuned calibration: less mind wandering when the clock hand approached the upright position and more mind wandering throughout the rest of the trial. These findings suggest that people aren't only aware of their minds wandering but also have some control over when they do it. Participants were on task when required and wandered further afield during the fifteen seconds or so of free time.[44]

Deciding when to let our minds wander into the future is an expression of metacognition—thinking about thinking. We selectively daydream about our upcoming beach vacation when we know we have the opportunity to settle in and enjoy the ride. Recall the bus journey from Chapter 1. As the vehicle nears your destination, you put aside thoughts about future tasks or the fantasies about palm trees and piña coladas so you make sure you don't miss your stop. This control we have over daydreams is in stark contrast to actual dreaming. In a dream, we simulate mental scenarios, not unlike when we imagine the future in our waking hours. However, we generally don't do this deliberately or catch ourselves dreaming. Some people do report having lucid dreams, where they do know they are off with the fairies, but for the most part people only become aware that their nighttime hallucinations were just dreams after the fact. It is upon awakening that we can look back and pull a Freud, pondering what a dream meant.

The ability to deliberately allow our waking minds to wander has been linked to the way the default mode network communicates with another key brain network, the *frontoparietal control network*. This network is active during controlled and effortful tasks like sustaining attention, keeping memory items in mind, and solving difficult problems.[45] In one study on mind wandering led by the clinical cognitive neuroscientist Claire O'Callaghan, people with dementia syndromes were more likely to think about stimuli in the world around them rather than drifting into thoughts about hypothetical worlds past and future. This shift to being more "stimulus bound" was associated with alterations in how strongly the default mode network and frontoparietal network were connected.[46]

A central hub of the frontoparietal control network is located in the prefrontal cortex, which has undergone rapid expansion in human evolution. The prefrontal cortex is implicated in executive control, such as determining when enough mind wandering has taken place and it is now time to put plans into action, reflecting on one's own cognitive abilities, or inhibiting competing demands in order to pursue long-term outcomes.[47] But while we have some measure of control over where we take our time machines, we have far less control over how the journeys make us feel.

———

A man is as much affected pleasurably or painfully by the image
of a thing past or future as by the image of a thing present.

—Baruch Spinoza (1677)

It's easy to overlook just how astonishing it is that the mind can fashion anticipated joy or agony from mere imagination. We simulate future events, and then use the feelings triggered by those simulations in the present to predict how it would feel if the events were to really happen.[48] In this way, the advanced gearwork of the human mental time machine piggybacks on the ancient biological systems of emotion, which enable us to evaluate whether objects, entities, and events in the world around us are good (and should be approached) or bad (and should be avoided).

Many lines of evidence suggest that mental imagery is treated similarly by the brain to real perception, even if—thankfully—we rarely confuse the two.[*49] For example, many of the same neural regions become active when people are shown, say, an apple, and when they imagine an apple—it's just that in the latter case the trigger comes from the inside rather than the outside.[50] Emotions follow suit. Imagining something joyful or frightening can have the same physiological effects as if the stimulus were right in front of us, leaving us with sweaty palms and a thumping heart. Imagined or forecasted emotions can be just as strong as the actual feeling.[51] In the German vernacular, anticipated joy is even said to be the greatest joy (*Vorfreude ist die schönste Freude*).

We have found in one of our studies that simply asking people to imagine emotional future events can encourage them to delay gratification.[52] Participants made decisions about rewards that were available either sooner or later. *Would you rather have ten dollars now or fifteen dollars in a month?* Before

---

* Interestingly, some people report having no mental imagery whatsoever: a complete inability to conjure up a mental picture of, say, an apple, car, or house. This reported lack of mental imagery has been called *aphantasia*, though these individuals may simply fall at the extreme tail end of a normal distribution of mental imagery intensity. There is some evidence to suggest that people who struggle to conjure up such mental imagery, perhaps unsurprisingly, report a lack of vivid sensory detail when imagining the future.

making each choice, they were prompted to imagine an emotionally charged event they might experience around the time of the delayed reward, such as a birthday party or getting sick. Imagining the future, whether positive or negative, led people to choose the later reward more often. In brain imaging experiments, this kind of shift towards delayed rewards relates to how much the hippocampus connects with regions of the prefrontal cortex involved in emotional evaluation.[53] These findings suggest that by vividly anticipating the impact of future rewards, the brain can assign them greater value, offsetting our natural tendency to discount the future.

But, of course, anticipated emotions can nonetheless be way off the mark when it comes to foretelling what the future will actually feel like. The social psychologist Daniel Gilbert and his colleagues have spent years tracking how our mental time machines might lead us astray, including via some consistent emotional biases. For one thing, people tend to expect that emotional reactions will be stronger, and last longer, than they actually do.[54] For example, people report feeling less positive than they predicted upon winning money or going out for a nice dinner, and the good feelings wear off faster than anticipated. Charles Darwin seems to have appreciated this quirk when he wrote about envy: "The wish for another man's property is perhaps as persistent a desire as any that can be named: but even in this case the satisfaction of actual possession is generally a weaker feeling than the desire."[55]

For negative emotions, this bias exists partly because people neglect to take into account how capable they'll be at coping with the downturns in life. For instance, when people are asked to rate how they would feel if their romantic relationship were to end, they predict feeling worse than they typically do if it actually happens. After the breakup, coping mechanisms kick in and the storytelling lets us pick up the pieces: *We were never really a good match anyway; there are plenty more fish in the sea.*[56] Humans are characteristically good at telling stories, especially to ourselves, which means it's easy for us to justify why things might not be that bad after all (or why we were right all along).

Ultimately, thinking that an outcome will be worse or better than it actually turns out to be might give us an advantage. As our collaborator

Beyon Miloyan has pointed out, it drives people to put more immediate concerns aside and focus on averting the bad or achieving the good.[57] Just think of deliberate practice. As we saw in the last chapter, there are many more immediately satisfying things one can do with one's time than effortfully grinding away at piano practice. But if you want to become a professional pianist, it may be motivating to imagine the immense pride that will sweep over you as you step out to bow in front of a cheering crowd at the end of your first concert.

The evolutionary logic of exaggerating how terrible it would be to, say, lose a limb, is likewise obvious: it spurs us on to do everything we can to prevent it from happening to us. In one of our studies, we found that this bias in predicting negative emotions starts young: even four-year-old children overestimate how bad it will make them feel to lose at a simple game.[58] When it comes to preparing for negative possibilities such as threats—before the enemy is at the gates or the house is really on fire—our tendency to exaggerate emotions can be a powerful motivator. Unfortunately, this is a double-edged sword. Most of us spend considerable time worrying about things that may never actually happen. As the Roman Stoic philosopher Seneca put it two thousand years ago: "Foresight, the greatest blessing humanity has been given, is transformed into a curse. . . . Memory brings back the agony of fear while foresight brings it on prematurely."[59]

Sometimes terrible prospects are triggered by something happening in our immediate environment, like picking up one's phone to see a dozen missed calls from a family member. At other times we simply wind up in a frightening place as our mind wanders. We can be lying safely in bed at the end of a fulfilling day, and before long our brain dredges up potential troubles: mulling over our finances, worrying about our loved ones overseas, or pondering the potential embarrassment of our underpreparedness for tomorrow's meeting. Here, instincts and imagination are knitted together: the brain's fear systems react to a simulated version of reality, letting us feel the emotional force of negative possibilities in a way that can overcome our other preoccupations. The human brain has been compared to a smoke detector, wired to err on the side of caution. This better-safe-than-sorry design is what makes us jump out of our skin at the sound of a door slamming or startle at the sight of a snake. It also explains why our brains are so prone to simulating a multitude of possible dangers.[60]

———

In a prescient book from 1943, *The Nature of Explanation*, philosopher and psychologist Kenneth Craik explained the functions of a scenario-building mind that can entertain multiple future possibilities: "If the organism carries a 'small-scale model' of external reality and of its own possible actions within its head, it is able to try out various alternatives . . . and in every way to react in a much fuller, safer, and more competent manner to the emergencies which face it."[61]

It is easy to explore a model built by your own brain right now. You can probably work out how many windows are in your house or apartment without having to look up from this page and without previously having that number at the ready. Try it. In your mind's eye, fly through the rooms and count the windows. Mental space travel, if you will. Similarly, you can also use this ability to travel in time and revisit where you once dropped a vase or plan where to put a fire extinguisher. By repeatedly imagining actions and consequences, say, picturing how we might respond to an emergency, we become more prepared. When we then actually encounter a relevant scenario in the real world, we already have a locked and loaded expectation of what's coming next—and what to do about it.

In a study led by the cognitive neuroscientist Roland Benoit, simply imagining interacting with liked or disliked people in a boring, unemotional location, such as an elevator, caused participants to eventually like or dislike the location. This transfer of value from the imagined people to the imagined places was underpinned by activity in regions of the prefrontal cortex that are associated with integrating emotion and valuation.[62] In another recent experiment, participants became less frightened of a tone that had previously been paired with an electric shock by simply imagining that tone without the shock fifteen times. Here, too, the prefrontal cortex was involved in the gradual change in how much people feared the sound.[63] Therapists use a similar process all the time to help people unlearn unwanted associations. In *systematic desensitization*, for instance, people who suffer from a phobia—say of spiders—are asked to imagine a spider as one of the first steps in helping them eventually cope with coming face-to-face with a real one.

This ability for purely mental learning unlocks a huge amount of potential because it allows us to adapt without suffering inconvenient real-life consequences. We can even learn from things that would otherwise kill us if we tried them out in reality. Though skating on thin ice might at first seem a good idea, taking a second to consider what could possibly go wrong can teach us to avoid the risk.*

The radical behaviorists had it wrong. By focusing entirely on observable stimulus and response relations, they missed what makes us so powerful.[64] Our brains allow us to conjure up models of the infinitesimally small, unfathomably large, or purely fictional—just think of quantum physicists, astronomers, and novelists. Without moving a muscle, we can review yesterday, consider tomorrow, and plot the way ahead.

————

There is still much we do not understand about the neurological basis of our mental time machines. Deep questions remain—about how to interpret the signals from neuroimaging studies, about how to connect our knowledge about the workings of single cells to that of specific regions and brain-wide networks, and especially about how any of the buzzing activity of physical matter can give rise to a conscious, thinking mind at all, let alone one that projects ahead by minutes, hours, days, or years.[65]

In fact, even the simpler acts of perception and learning that we discussed at the start of this chapter are harder scientific problems to crack than they might at first appear. Our perception of the world around us is so seamless, so intimate, that it is easy to overlook just how much is going on under the hood. There is an immense computational dance happening in every moment between incoming sensory data and existing models of the world.

---

* Thoughts about possible events that might kill us can have a peculiar attraction. Upon noticing that a small step over a ledge or a minor twist of one's steering wheel could end it all, even people who otherwise have no suicidal tendencies may be drawn to entertain how those scenarios might play out. The French call this sensation *l'appel du vide*, the call of the void.

Charting how this dance unfolds through the electrochemistry of brain cells remains a major challenge for cognitive neuroscience.

The data we do have are converging on a view that predictive processes are part and parcel of how brains function across the animal kingdom. We have seen that brains use prediction not only to anticipate the future but to learn from the past and perceive the present. Brains are model makers, intimately familiar with their surroundings in a way that enables navigation towards states that maximize flourishing and away from those that threaten peril. But many of the abilities we have encountered in this chapter—weaving a narrative about possible futures, having emotional reactions about potential events, learning about future possibilities purely in the mind's eye, building a sense of self through time, even becoming aware of one's mortality—may well be the distinct purview of the brain of our peculiar species. Are we, in fact, alone on this planet in our mental access to the fourth dimension?

CHAPTER 5

ARE OTHER ANIMALS STUCK IN THE PRESENT?

Wild animals run from the dangers they actually see, and once they have escaped them worry no more. We however are tormented alike by what is past and what is to come.

—Seneca (ca. 65 AD)

For decades, Old Tom had helped the Davidson family in the little coastal township of Eden, Australia. At the beginning of winter, he would arrive from down south and alert the Davidsons' whaling station when his group had herded a baleen whale into Twofold Bay. Then the whalers would row out and kill the whale. It was a mutually beneficial arrangement: the Davidsons obtained the valuable meat, oil, and bones, while Old Tom and his mates only expected a small part of the catch, a delicacy. After Old Tom passed away in 1930, a whole museum was dedicated to him. To this day, it houses his hefty twenty-three-foot skeleton. Old Tom had somehow learned that breaching with his six-ton body and thrashing with his tail outside the Davidsons' whaling station would lead the humans in their rowing boats to help his pod kill the large prey whose tongue and lips they craved. The whalers would leave the carcass hitched to their boat until Old Tom and the rest of the orcas had their feed.[*1]

Are we really alone in thinking about the future, as Seneca assumed? Stories of smart animals like orcas, orangutans, and crows suggest otherwise. Humans are very brainy of course, but we do not have the largest brains on Earth. Whale brains can measure a whopping sixteen pounds, whereas ours weigh in at around three pounds. But whales also have the biggest bodies more generally, so perhaps it is more appropriate to compare relative brain sizes. Our brains make up about 2 percent of our overall body mass, whereas the brains of whales constitute less than 1 percent of their body mass. Other creatures, however, have much larger relative brains sizes than humans. Some top-heavy shrews have brains that make up about 10 percent of their body.

Since we get beaten by large mammals in one comparison and small mammals in the other, researchers have proposed a third comparison, the so-called Encephalization Quotient, which takes into account that as mammals get bigger, brains get absolutely larger but relatively smaller as a proportion of their body. In this scheme, humans do come out on top, even if one may well be suspicious about whether we are just using statistics to confirm our presumptions. We have over seven times as much brain mass as would be expected of an average mammal of our size. Dolphins are not far behind, with some species equipped with a brain more than four times the mass expected for a mammal of their size. Dolphins are widely assumed to be exceedingly smart, and orcas, in spite of also being known as killer whales, are actually the largest species of dolphin. Perhaps dolphins are the nonhuman animals most likely to be capable of thinking ahead like humans.[2]

Dolphin foresight is not easily studied in their wild aquatic habitats. But some tantalizing observations have been reported. In addition to cooperating with humans to catch prey, like Old Tom did, pods of orcas have been

---

* Orcas naturally hunt migrating baleen whales. For untold generations, the Indigenous Thaua people of the Yuin nation benefited from orcas herding the whales into Twofold Bay, where some would be stranded. When Thaua people later crewed on European whaling boats, they would let the revered orcas feast on the tongues of harpooned prey. The Davidsons absorbed this approach instead of driving the orcas away as other whalers did. Eventually, the orcas learned to assist the whaling family.

Figure 5.1. Old Tom swimming with a calf alongside a whaling boat towed by a harpooned whale.

seen making waves to sweep stranded seals off drifting ice rafts. Dolphins blow air bubbles to herd fish into convenient balls that they take turns feeding upon, and they sometimes carry sponges over their beaks as they scour the sea floor, apparently to prevent them from cutting themselves on sharp rocks or on the barbs of stingrays.[3] There are other intriguing anecdotes, such as orcas using bits of fish as bait to lure seagulls within their biting range, but it is difficult to disentangle what actions the cetaceans are really planning, and which ones, by dint of circumstance, simply wind up being useful.[4] For this, we need well-controlled scientific studies.

At Disney World in Florida, scientists gave two bottlenose dolphins, Bob and Toby, a series of puzzles to put their planning abilities to the test.[5] In one task, the dolphins were first shown how to pick up weighted rings with their beak and were then taught how to drop four of these weights into a container to release a tasty fish reward. Once Bob and Toby had learned this, the experimenters changed the difficulty level: now weights were spread out within 20 feet of the prize box, meaning the cetaceans had to do some fin work. Yet rather than simply gathering up the four required weights in one trip, they did it by swimming back and forth between each object and the container. After several dozen trials, the experimenters moved the weights farther away still from the prize box, within a radius of nearly 150 feet. Now this was too much hard work: gradually, Bob and Toby started putting multiple weights on their beak at a time to shorten the journey. They did not simply pick up the required four, however. Often the dolphins gathered up two, three, and even five weights, so they may have struggled a bit with the counting.[6] Nonetheless, by carrying more than one weight at a time, they demonstrated at least some capacity to think ahead.

In another task, the dolphins had to put a single weight into a box, but this time they also had to poke a stick inside to obtain the reward. There was a twist. The second step was only possible for about fifteen seconds after the weight had been dropped, before a sliding door sprang shut and prevented further access. This did not pose a big problem for the dolphins: they quickly learned to complete the sequence in time. Next, however, the researchers placed the stick over 80 feet away from the box. Bob and Toby once again dropped the weight into the apparatus and then quickly swam to retrieve the stick. But when the door kept shutting before they had returned, they simply gave up. With a little more foresight, a simple solution would have presented itself: go and get the stick first, glide back at a leisurely pace, and *only then*, with the stick at the ready, put the weight into the box. The dolphins did not get it. They did not prepare.

Dolphins can evidently plan to some extent, but their enduring errors, even after many trials and opportunities to learn, suggest their foresight is quite restricted. As we will see, such results are typical for studies of animal planning. On the one hand, there is evidence of some competencies. Animals are not just mindless automatons. On the other hand, performance tends to be inconsistent, and tasks that might seem trivial to a human mind, even to a young child, often go unsolved.

Humans hold rather conflicting ideas about the minds of other animals. Some people are attracted to what we call *rich* interpretations of animal behavior and readily attribute complex cognitive capacities to animals, while others are reluctant to do so and instead gravitate to *lean* interpretations.[7] Many people vacillate between these views depending on the context (and what's on the menu that evening). On the one hand, people frequently anthropomorphize, projecting all kinds of mental processes onto their pets—feelings, memories, expectations. On the other hand, those same people may treat other animals, especially those farmed for food, as if they had no mind at all.

Scientists, not immune to holding preconceived ideas, should guard against any biases influencing their research. Sensational rich claims about animals apparently thinking ahead in clever ways can be exciting, but they cannot be simply accepted at face value. These claims need to be tested in rigorously designed studies, and results need to be independently replicated. Before jumping to any conclusions about animal planning capacities, we must systematically rule out lean alternative explanations. As we will see, an animal may have just repeated what was previously rewarded or may have acted on instincts that seem clever and farsighted without any real understanding of what the future holds. In this chapter we will find out what science has thus far established about foresight in other species.

Animals can give the impression of being focused on the here and now, such as when a lemur basks when the sun appears from behind the clouds.

But this does not mean they are completely stuck in the present. Most species, great and small, face recurring patterns in nature such as fluctuations of light, temperature, and food availability that occur in periodic cycles. Even the humble bacterium *E. coli*, infamously responsible for food poisoning, prepares. As it travels through lactose-rich human digestive tracts, it switches on genes for maltose digestion a couple of hours before it will reach the maltose-rich areas. This preparation does not mean that the bacterium is fantasizing about maltose, however. The strains of *E. coli* that happened to activate genes in this order survived and replicated more than the ones that did not, or the ones who did so too early or too late. If the long-term pattern stays the same, like how maltose always comes after lactose in the digestive tract of a host mammal, natural selection can forge behaviors that seem intelligently calibrated to upcoming events. The key takeaway here is that only genetic variability and a reliable sequence of environmental circumstances are required for such forms of preparation to evolve.[8]

Creatures that act in tune with long-term regularities such as daily or seasonal variations can have a significant advantage over those that do not. There is perhaps no more conspicuous a case of preparation than that of squirrels and other animals storing food for barren winter months ahead. One would be forgiven for assuming that the squirrels must be imagining themselves hungry and without food in the midst of the coming frost. But this is not why they hoard food. Even a young squirrel that has never experienced a winter will collect and store provisions. This simple fact tells us that the behavior is driven by instinct rather than insight. In other words, the squirrels have evolved a behavioral solution to the recurring challenge of wintertime food shortages.

Figure 5.2. Living in the moment: A lemur apparently focused on the warmth of the sun's rays.

In a sense this adaptation may not be that different from whales storing fat in their blubber for migration without food, or Australian trees storing energy in swollen lignotubers at the base of their trunk that they can draw on after a fire. So animals may end up preparing for nightfall or winter even if they do not think about the approaching darkness or cold. Though reliable, the mechanisms underlying these behaviors tend to be relatively inflexible and may not be so helpful in the face of new challenges.

Such a lack of flexibility is illustrated by the classic example of the greylag goose's response to an egg rolling away from her nest. First, she will rise to her feet, extend her neck over the egg, and then carefully begin to roll it using the underside of her bill. She will then push the egg towards her legs and walk slowly backward into her nest, all the while faintly moving her head from side to side to balance the egg and ensure it doesn't escape her clutches. The mother goose's behavior of course has an important future-oriented function: it prevents her from accidentally breaking her egg and eliminating a valuable opportunity to pass on her genes. But this does not mean she has thoughts of a precious little gosling on her mind as she approaches and rolls the egg. If a mischievous experimenter distracts the doting mother and removes the egg from under her, she will nonetheless continue the series of actions to completion. In fact, when Nobel Prize winners Konrad Lorenz and Nikolaas Tinbergen placed other egg-shaped objects near the nest of a mother goose, she behaved in exactly the same manner. She even carefully rolled objects that do not resemble an egg much at all, such as cube-shaped toys.[9] The mother goose appeared to be trapped in a behavioral program that was automatically elicited by certain stimuli.

It's not just Nobel laureates who can play tricks on mother birds. Cuckoos lay their eggs in other birds' nests, and, upon hatching, the cuckoo chick's first act is often to destroy the host's own eggs. Afterwards, the deceived mother bird is helplessly drawn to putting food into the intruding chick's mouth, even if the chick is bursting out of the nest at several times the mother's size. In this case, the cuckoo chick is exploiting the fixed tendency of the mother bird to feed a gaping, chirping red mouth in her nest—apparently with no regard as to who that mouth belongs to.

Figure 5.3. A fan-tailed cuckoo chick (*left*) is fed by a female scrubwren (*right*) trapped in a fixed behavioral pattern at Samsonvale in Queensland, Australia.

In a sense, the cuckoos and the hosts they dupe are in an evolutionary arms race.[10] There is selection pressure for the potential host to detect and reject brood parasites—if a reed warbler sees a cuckoo, for instance, it is more likely to desert its current nest.[11] In turn, some cuckoos emit a hawklike chuckle after laying eggs in another bird's nest, which has the effect of distracting the host and thus increasing the chances of getting away with the cuckolding. All this complex behavior is evidently future directed in nature, but it can run along without anyone having formed a considered plan.

Life is dangerous. Animals that predict threats can increase their chances of avoiding harm to themselves or their offspring by taking preemptive action. When chickens see a moving shadow on the ground, they often respond by looking up, apparently to check for potential danger such as a hawk flying overhead. Like cuckoos and their hosts, predator and prey are also pitted against each other in an evolutionary arms race, favoring the early detection of threats and hunting opportunities.[*12]

Different species are vulnerable to different predation risks, and so they have evolved diverse ways of detecting and dealing with these threats. As they graze, gazelles tend to scan the horizon, which increases their chances of spotting a potential terrestrial attack by predators such as large cats, whereas many monkeys also keep an eye on the skies because they are vulnerable to dive-bombing birds of prey. Animals become more vigilant in situations that are particularly dangerous. Monkeys who are active during the day are more cautious at night, while nocturnal animals such as rats are more cautious in the light of day. Grazing animals tend to be more cautious in open spaces, but less so in large groups—presumably because the chances of falling prey are reduced and the chances of detecting a threat are increased with more eyes on the lookout.

Animals can be threatened by predators, by microbes that cause illness, by a lack of vital resources such as food and water, and by environmental hazards such as raging fire, floods, or storms. In social species, conflict can lead to ostracism, and even failing to find a sexual partner is threatening in the sense that it entails a hard stop to a genetic lineage.[13] Given the range and reliability of recurring dangers, it is not surprising that species have evolved means of dealing with these situations. These behaviors may look superficially like the result of foresight, but they are primarily instinctual.

———

Of course, animals do not just execute fixed behavioral programs. Associative learning allows them to establish expectations about the particular threats and opportunities in their local environments.[14] As we saw in Chapter 1, when Ivan Pavlov found that dogs associated a ringing bell with the

---

* When a predator is on the hunt, its prey typically has three types of response options. First, the prey animal can hide or stay perfectly still to avoid drawing any attention. The second option is to get away from the danger as fast as possible: run, fly, burrow, or climb away to safety. Finally, it can prepare for battle: get defenses ready for the struggle ahead. Just imagine you are swimming off the beach and suddenly see a huge dorsal fin cut through the waters in front of you. You could stay still and hope for the best, swim away as fast as you can, or ready yourself to punch and kick. *Freeze, flight,* or *fight.*

arrival of food, the animals were really learning what *predicted* the arrival of the treat and preparing by salivating.[15] In another experimental paradigm, researchers in the 1960s would put a dog in a cage with two compartments, briefly dim a light, and then electrify one of the compartment floors. The dog would quickly jump into the other compartment to escape the cruel shock. Subsequently, the dog avoided the shock by jumping into the other compartment as soon as the light dimmed. So here it learned to use a cue to predict something unpleasant and avoid it.[16]

For much of the twentieth century, behaviorists stripped back the environment of their animal subjects—usually rats or pigeons—and controlled what stimuli they experienced and what reinforcement would be available. B. F. Skinner championed this era of research with studies on animals in boxes that came to be known by his name. When the animal behaved in a particular way in a Skinner box, say it pecked on a disk when the light turned green, it would be rewarded with food, whereas other behaviors might be punished. Animals readily learn which action predicts which outcome in which context.[*][17] And human trainers have exploited these abilities to put animals to work, from getting horses to execute a precise series of movements in a dressage competition to teaching dogs to sniff out drugs at the airport (even if Skinner's idea of using pigeon-guided missiles during World War II did not take off).

We now know that animals not only learn that an action is rewarded but also form a specific expectation about the type of reward itself. In one study, capuchin monkeys who had learned to hand an experimenter a pebble in exchange for a piece of cucumber refused to continue the task once they saw another monkey receive a more appetizing grape for performing the same action. The cucumber was no longer good enough. The scientific article describing these results was titled "Monkeys Reject Unequal Pay," and this rich interpretation has become a popular meme.[18] However, there is a leaner

---

* The effects of systematic rewards and punishments are strikingly reliable. The most intense learning occurs when it is not exactly predictable how many actions are required to bring about the reward. This is one reason slot machines are so addictive—you get rewarded every now and then, so long as you keep adding money. Technically, this is known as a variable ratio reinforcement schedule.

interpretation that does not entail any desire for equal pay. The monkeys also tend to reject the cucumber once grapes are simply put in sight but out of reach. Chimpanzees have been similarly tested, and they too stop such exchanges when they suspect that the experimenters are not rewarding them as well as they could be—even without another animal present. This is not about inequality, then, but about someone not giving you what you want.[19] Still, these results do show that primates can learn to expect specific kinds of rewards—and evidently what they expect can change quickly.[20]

Associative learning can occur independent of any awareness of true cause and effect. In a classic study, Skinner rewarded pigeons every fifteen seconds—completely irrespective of their actions—and they eventually displayed peculiar and idiosyncratic behaviors.[21] One pigeon would make counterclockwise turns; another would toss its head upward. The explanation is simple. Whatever behavior the animal happened to be doing just before the reward arrived was reinforced and therefore likely to occur again. This is a self-perpetuating process because the behavior is more likely to occur and be reinforced again at the end of the next fifteen-second interval . . . and the next. Similarly, when people's actions are followed by reward, they tend to repeat them next time around. Athletes may kiss the badge, wear that charm, or go through any number of odd rituals before they start the competition. When asked, many people admit that there is no causal reason, but under pressure they may still pick those lucky socks.

So although prediction and expectations are essential to associative learning, this does not necessarily entail that animals are aware of causal relationships. Nor does it imply that they ponder remote future events. A delay of only minutes typically makes the learning of associations between events impossible.[22]

With that said, at least some animals have long-term memory from which to learn for the future. They can store some kinds of information about *what* is *where* to guide their future actions. It has even been argued that they store information about *when* something occurred. The comparative psychologist Nicola Clayton and her colleagues have found scrub jays, birds with an instinct to cache foods, can learn to adjust their search for hidden morsels depending on how long ago they stored them.[23] If they have cached both mealworms and nuts only four hours ago, the birds will search for where the worms were hidden because worms are their preferred food. By contrast, if they have cached both mealworms and nuts five days ago, they will instead search for nuts—as the mealworms would be spoiled by this time. The scrub jays seem to know what is where, and whether it is still good to eat. This may be because they recall the episodes when they stored the foods. But it need not. It could also reflect associative learning. Memories tend to fade with time, and the birds may learn to associate the strength of their memory for a food's location with how good it tastes when retrieved. Worms are no good once the memory has faded even just a little, so digging up worms is only reinforced when the bird has a strong memory for their location. By contrast, digging up nuts, even if the memory of their whereabouts is very weak, is rewarding every time. Such relatively simple learning can explain the behavior without a need to assume the birds are mentally revisiting when they cached the food.[24]

You may well wonder if it could ever be incontrovertibly established whether animals do or do not have episodic memory.*[25] How could we know if they have anything like the recollective experience we humans have when remembering our past?[26] At the end of the day, all we can directly see is animal behavior.

But given that mental time travel into past and future are linked in mind and brain, and given the fact that foresight is so clearly adaptive, if animals had mental time travel capacities, we should expect to witness their foresight in action.[27]

———

By thinking ahead and comparing possibilities, humans realize that the world is full of mutually exclusive trade-offs. We cannot have our cake and

---

* In humans, knowing what happened where and when need not entail mental time travel into the past. After all, we can know such information about many events—just think of, say, your own birth or even historic events—without being able to mentally reexperience them. And conversely, much of our episodic memory tends to be factually wrong about details such as when and where something exactly occurred—*Did that happen on New Year's Eve two years ago? Or three? At my place or yours?*

eat it too. Eating food means it cannot be stored, and storing food means it cannot be eaten now (and comes with the risk that it will spoil or be stolen). As we saw in Chapter 3, one of the keys to smart choices about the future is not merely being able to delay gratification per se, but dexterity in deciding when to wait and when to just go for it: flexible decision-making that accounts for diverse objectives, opportunities, threats, and possibilities. Do other animals simulate alternative possibilities? Do they compare immediate and delayed options, weigh up their risks and benefits, and act accordingly?

Research has established that some animals will delay gratification, at least when an anticipated food reward is very attractive and the delay very short. Pigeons and rats may pick delayed larger rewards over more immediate smaller rewards if the delay is a few seconds in duration.[28] Monkeys will sometimes wait many seconds.[29] Our closest animal relatives, chimpanzees, can wait several minutes holding on to a small reward without eating it until they can exchange it for a much larger reward, or when rewards gradually accumulate as long as they do not reach out and grab any.[30] But there is no experimental evidence yet showing anything remotely like the human capacity to deliberately delay all manner of potential rewards for days, years— even lifetimes. We study hard to learn skills that will only pay off in decades to come, and we restrain our spending for a rainy day, for retirement, or even to pass on a nest egg to our descendants.

Despite decades of dedicated observation, we still have very few signs of apparent foresight in the natural behavior of our closest animal relatives. After observing a male chimpanzee building a sleeping nest close to a female, Jane Goodall pondered whether he may have been planning an opportunity for mating the next morning.[31] Similarly, chimpanzees have been found to prefer building nests in places that have attractive breakfast options nearby.[32] But it is difficult to establish any future-directed intent in such cases. After all, current drives can explain why a chimpanzee might want to be close to something desirable. One study even claimed that adult male orangutans inform others about where they are headed by making loud calls in that direction.[33] However, here too there are leaner interpretations that are difficult to rule out. After all, orangutans tend to face both in the direction they are heading and in the direction they are calling, whether or not they are intending to inform others about some travel plan.

As we saw in the last chapter, cells in the medial temporal lobes work together like a mental map, enabling mammals to move towards potential positive events and away from potential negative events.[34] Single-cell recording has even shown that hippocampal cells of rodents fire in particular sequences not only when moving through a maze but also later on, when the animal is out of the maze and running on a treadmill or just resting. This might suggest the animals are mentally replaying past movements through space or even preplaying potential future routes. These neuroscientific data led Michael Corballis—who, in collaboration with Thomas, had introduced the concept of mental time travel and argued for its unique role in human evolution—to change his mind: maybe rats can travel mentally in time after all. Unfortunately, the exact functions of these neural patterns have remained unclear, let alone how they relate to any mental experience or the capacity to anticipate future events.[35]

Each sequence of rodent hippocampal replay typically takes only a fraction of a second, which would make for some very rapid journeys through time. Mental time travel, as humans know it, often takes a lot longer. You can play out lengthy past and potential future scenarios and envision these scenarios occurring in roughly real time, as you might do while imagining how to score a game-winning goal while currently standing in the shower.

It is nonetheless intriguing that, when rodents are in a maze, hippocampal firing can rapidly alternate between representing the left and right sides of an upcoming fork, and often indicates which direction the rodent will choose. To help shed light on what might be going on here, try to imagine you are approaching a green traffic light in your car when suddenly it turns yellow right at the awkward moment when you just might be able to enter the intersection before it turns red. Do you hit the brakes, or do you floor it? Quick! In scenarios like these, we may find ourselves *caught in two minds* about what to do, and we may even find our foot hovering between the brake and gas pedals as we feel the simultaneous forces of the two different options competing to control our behavior. Afterwards, perhaps with a sigh of relief, we integrate these two mutually exclusive courses of action into a single event (*Lucky I hit the brakes; that blue car might have T-boned me if I had gone through*).

Of course, humans can integrate future possibilities not only after the fact but also long before the heat of the moment is upon us. If your favorite band is playing their last show on the same night as your grandma's eightieth birthday celebration, then you know very well that choosing to attend only one of these occasions will forever close the door on attending the other. By contrast, rodents may simply get caught in two minds about which is the better side of the forked maze and never understand that these two possibilities are mutually exclusive, as humans readily could.

In the wild, several species of weaver birds have an instinct to build nests with two entrances, which ornithologists suggest have a peculiar function: "If a predator started to enter one, the occupant could fly out the other."[36] But human hunters would readily lay a trap in front of all of a prey animal's potential escape routes out of such a hiding spot, rather than just in front of one, because they foresee mutually exclusive possibilities: they know the target could emerge from any exit.

Figure 5.4. A grey-capped social weaver nest with two holes at the bottom.

Recall from Chapter 1 that we examined something quite similar with the forked-tube task. We did not just test children but also nonhuman primates. First, we showed chimpanzees and orangutans a vertical tube and dropped a reward into a hole at the top so they could catch it when it fell from the bottom.[37] After watching this only a few times, all of the apes readily anticipated that the reward would reappear at the bottom of the tube: they placed their hand under the exit to prepare for the catch. Next, we made the future a little harder to predict. We replaced the straight tube with the forked tube that had two exits at the bottom, kind of like the weaver nest.

As you might remember, by age four the vast majority of children immediately and consistently secured their reward by covering both exits with their hands. By contrast, all of the apes failed to demonstrate this simple form of contingency planning. They tended to cover only one of the potential exits when preparing for the drop, and therefore caught the reward only around half of the time. Even after many additional trials, they did not seem to understand that a single event can have two mutually exclusive outcomes.

In subsequent studies we created new versions of the problem and compared the children's performance with that of apes and monkeys. These versions included a forked tube made out of clear plexiglass so that the path of the reward remained visible, and two parallel tubes where we simply held the reward in the center before rapidly dropping it into one of them.[38] No matter how we designed the task, we kept finding that children's performance gradually improved with age, whereas the other primates of all ages tended to cover only one of the exits. On rare occasion, a few of the primates covered both exits, demonstrating that they have the physical capacity to solve the task. However, they all subsequently went back to merely covering a single exit.[39] Their problem was not in their hands but in their minds. The penny didn't drop.[40]

Although there is much more work to be done on this topic, there is as yet no compelling evidence that nonhuman animals, even our closest living great ape relatives, can foresee mutually exclusive possibilities and prepare accordingly. This may indicate a profound limit to animal minds. Without

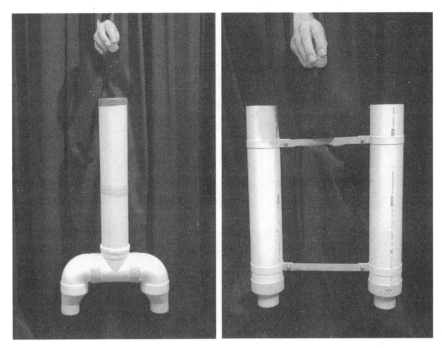

Figure 5.5. The original forked-tube task (*left*) and a version where the experimenter drops a target into either one of two parallel tubes (*right*).

———

such a capacity, they might not be able to plan for contingencies, weigh up alternative courses of action, or lament their mistakes.

———

Chimpanzees are undoubtedly smart, and it is clear that they can think ahead in at least a basic manner, just like dolphins can. If we ever drop a grape by accident outside their zoo enclosure, the chimpanzees we work with will quickly retrieve a stick and use it to roll the treat within reach. They will also sometimes walk away to make a suitable tool by stripping off leaves or breaking twigs in half, even if they can no longer see the grape, which suggests they can maintain the problem in their mind's eye.[41] In their natural habitat, chimpanzees use many tools, such as stones to crack open nuts. And they will sometimes even pick up a useful hammerstone a hundred yards or more before reaching a tree where they will use the stone to crack the nuts.[42]

Figure 5.6. The chimpanzees Cassie and Holly at Rockhampton Zoo have been prolific participants in our studies.

At Furuvik Zoo in Sweden, one male chimpanzee called Santino was observed piling up stones and bits of concrete in the morning, before later using them as ammunition to hurl at zoo visitors.[43] Santino seemed calm when he collected the ammo but was quite agitated—hair standing on end—when he later chucked the debris at the arriving visitors. This headline-grabbing anecdote suggested that the ape may have anticipated his future agitation and desire for stuff to throw—and prepared accordingly.

For humans, this kind of anticipation is par for the course. We regularly prepare for changing internal states, such as buying orange juice for tomorrow's breakfast even if we are not at all thirsty now, or bringing something warm to wear before it gets cold. The psychologists Norbert Bischof and Doris Bischof-Köhler proposed that this ability to imagine future motivational states—even when they conflict with our current states—might be what makes human foresight unique.[44] Many animals secure future needs, of course, such as when they cache food or build a nest, but this may not be done with any future motivation in mind. If Santino foresaw his own future state of agitation and therefore stockpiled things to satisfy his later desire to hurl objects, then this would run counter to the Bischof-Köhler hypothesis. However, such preparation has, as yet, not been documented in other zoos or in the wild, highlighting the need for controlled studies to assess whether other factors might have been driving this curious behavior.[45]

What we do have are a handful of studies that have given great apes the opportunity to secure a tool they would need to satisfy an appetite they probably already have.[46] Perhaps the most famous of these studies was published in *Science*, one of the world's very top journals, and it caused quite a stir.[47] Bonobos and orangutans were first trained to obtain grapes with a plastic tool from a feeding apparatus. They were then ushered out of the testing room, and the experimenters removed all the remaining tools before allowing the apes to return one hour later. On almost half of the trials, the apes carried a tool with them from one room to the other and back again—allowing them to operate the apparatus and obtain more treats upon return. One bonobo and one orangutan even returned with their tools after an overnight delay. These apes may have planned ahead, but because it was always the same tool that had to be selected, it remains unclear whether they really thought ahead to the future situation and de-cided to bring back the tool, or whether they had merely learned to associ-ate the tool with rewards (much like a favorite toy that a child might carry around because it has been associated with comfort in the past).

Since then, claims of amazing animal foresight have occasionally surfaced with considerable fanfare. But alas, simple alternative explanations have con-tinued to cast doubt on rich interpretations. In 2017, another study pub-lished in *Science* argued that ravens can plan for future events. The evidence came from five birds that had learned to pick up a stone and drop it into a box to get a reward.[48] Later, these ravens picked the stone from amid alterna-tive distractor items several minutes or even seventeen hours before the box was again made available. The researchers concluded that the ravens were flexibly planning ahead.

However, this behavior is only impressive at first sight. Before the tests started, the birds had learned, over several trials, that the stone could be used to retrieve rewards and that distractor items could not. It is possible, and indeed likely, that when the experimental trials began, the ravens therefore selected the stone not in anticipation of any future situation but because it had high value: it had already been associated with food. In fact, the birds had not been given any information about when the reward opportunity would return—the seventeen-hour interval was completely arbitrary and the birds had no way of keeping time. So, on the very first trial, all that these ravens could have possibly relied upon was the past association.[49]

Ravens and other corvids are often said to be clever, and one species in particular, the New Caledonian crow, has indeed shown competence at solving diverse problems, as we saw earlier when one made hooks out of wire. In the wild, these crows can be observed making tools from sticks and leaves, including hooked twigs they use to forage for grubs from inside crevices. Comparative psychologists Alex Taylor, Russell Gray, and

Figure 5.7. A New Caledonian crow using a tool to retrieve a reward from an experimental apparatus.

their colleagues have extensively examined the birds' cognitive capacities.[50] Having written many critiques of studies claiming to have shown animal foresight, Thomas recently teamed up with these researchers, as well as with Nicola Clayton and others who had previously advanced rich interpretations, in a study on the potential planning capacities of New Caledonian crows.[51]

Before being tested, the birds were trained to use different tools to get food from three distinct contraptions—a stick to poke into a tube, a stone to drop onto a platform, and a hook to operate a dispenser. In the main test, the crows were briefly shown which contraption was in a room. Then, five minutes later, they were presented with a selection of tools in a second room. Once they picked a tool, they had to wait another ten minutes before they were given access to the room with the contraption. If they had been initially shown the tube, for instance, they had to pick the stick so that later they could use it to poke out the reward in the other room. Given that the contraptions were changed between trials, the same tools were correct on some occasions and incorrect on others. Three out of four New Caledonian crows that had passed control conditions managed to select specific tools according to the situation they had good reason to expect. This not only demonstrates that the crows can pick something that has previously turned out to be useful but suggests that they can also pick something that will be useful for a specific future event.

Many questions remain, of course. What would the birds do if we visibly removed or destroyed the contraption before they had the opportunity to select the tool? Would they change their choice and, say, pick a low-value food item instead of the now-useless tool? More research is needed to identify what drives the crows' behavior and where the limits of their competences lie. But these positive findings do raise the possibility that richer interpretations of animals' behavior may not be unfounded.

The great apes in particular appear to show some basic capacity to think ahead, whether an experimental task involves transporting tokens, twisting paddles, or preparing tools.[52] Nonetheless, their foresight often falls short when the tasks are made just slightly more difficult. Ape behavior is not unlike that of Bob and Toby, the dolphins we met at the start of this chapter.

Yes, they may plan in some basic sense, but even after much experience, they act far from optimally and often fail to come up with even simple two-step solutions when required. The animal foresight demonstrated thus far remains quite limited compared to human abilities, and even animals' successes tend to involve a lot of errors.

———

A pattern of capacities that fall short when tasks are made more complex has also been found in other domains often considered the hallmarks of human cognition. In his previous book, *The Gap: The Science of What Separates Us from Other Animals*, Thomas examined the evidence regarding the most common factors often assumed to set us fundamentally apart from the rest: language, intelligence, reasoning about other minds, culture, morality, and mental time travel.[53] On close inspection it became clear that various species, in particular the great apes, have some capacities in every one of these domains. For instance, chimpanzees can solve problems through insight, console others in distress, and, as we saw earlier, even maintain social traditions. Nonetheless, careful examination of the available data also points to what is special about the human mind.

Two underlying characteristics keep reemerging as critical in all these contexts. First, *nested scenario building*, our ability to imagine alternative situations, reflect on them, and embed them into larger narratives. Nested scenario building refers to the ability we have been explaining here with the theater metaphor. So admittedly, it is not really one ability per se but an emergent property of interacting components. This mental theater allows us to imagine events from another time, and it also plays a key role in distinctly human forms of communication, problem-solving, mind reading, culture, and morality. For instance, we can use our imagination to walk in another's shoes, to simulate how that person might feel, think, or weigh up options.

The second factor is the *urge to connect* our minds together and exchange such mental scenarios. We have a deep-felt need to understand and be understood, and our open-ended language may have evolved primarily for broadcasting our internal plays. We use language to tell each other what we have experienced and to share our plans about the future. We ask questions and volunteer feedback about the episodes we recall or envisage. We tell stories and learn from the narratives of past, future, or fictional events that others share. These exchanges allow us to make better predictions and to coordinate our actions to shape the future increasingly to our own design. After all, the best way to find out about the future is to ask someone who has already been there . . . so to speak. If you really want to know what a holiday in Australia is like or what a career as a commercial pilot entails, you can envision all the fanciful scenarios you want, but your best bet is to chat with someone in the know—like a person who has visited Sydney or flown the plane there herself. So far, we don't have any evidence that other animals use their communication systems to go on shared walks down memory lane or collectively plot for a faraway future event.

As it stands, the evidence we have points to something distinct about the human capacity to construct mental scenarios, and to share them. More work is needed to precisely determine the foresight capacities of other animals as well as where their limits lie. As is the case with young children, nonhuman animals may be restricted by deficiencies in the components involved in the metaphorical theater production, and different species may be restricted in different ways.

Recall that to mentally entertain events and examine different mental scenarios one needs something akin to a stage: sufficient working-memory capacity to relate several chunks of information. Though there is no clearcut nonverbal test to measure it, the working-memory capacity of great apes has been estimated as two or three bits, which might severely limit what relationships they could entertain.[*54] Predicting how objects and actors behave on this mental stage, furthermore, requires some understanding of rules governing the physical and social world. While there are some intriguing

---

* Chimpanzees have performed impressively on touch-screen memory tasks. However, these tasks are not equivalent to standard working-memory measures, which tend to involve mentally manipulating information. The big-ticket finding that the chimpanzee Ai outperformed adult humans was quickly challenged by research showing that when humans are given considerable practice (as the chimpanzees had been), they can match the performance of the chimpanzees.

signs of competencies, much research suggests that nonhuman animals are severely restricted in their reasoning about aspects of our world they cannot directly observe, such as what goes on in the mind of another individual or what physical forces ultimately govern the movement of objects.*[55]

The playwright of the theater production refers to our capacity to combine and recombine finite basic elements into virtually endless novel assemblies. There is little to suggest that other animals have this generative capacity, whether in the context of imagining diverse future possibilities or in other arenas such as language, numbers, or music.[56] It is nonetheless possible, and perhaps even likely, that other animals have mental flashbacks of past events or even daydream—various mammals appear to have something like a default mode network, after all.[57] But to understand such mental experiences and appreciate them as part of a broader story, an animal would need to have at least the basic component capacity of what we called the director. They need to recognize that their mental scenarios are *about* a past or potential future event and evaluate them.[58] There is as yet no indication that nonhuman animals can think such nested thoughts.

For evolution to select for mental time travel, the mental activity must influence an animal's behavior in ways that somehow increase their chances of survival and the propagation of their genes. Thought must be put into action by something like the executive producer. To take full advantage of mental scenario construction, at least some executive skills are needed to inhibit more immediate urges and drive actions aimed at longer-term goals. Inhibiting temptations and scheduling one subgoal before another—such as when the dolphins needed to stop themselves from dropping a weight into a container until they had retrieved a stick from farther afield—is not a trivial demand. Recall that in simple delay-of-gratification tasks, animals only manage to resist temptation for relatively brief periods of time.

---

* Some species may, for instance, also have the foundations for a "theory of mind" that would allow them to attribute mental states—like desires and beliefs—to other individuals. In one promising line of research, a team led by comparative psychologists Christopher Krupenye and Fumihiro Kano used infrared eye tracking and found that, as a person prepares to search for a hidden object, great apes look at the location where the person falsely believes the object is hidden, even when the apes themselves know it is somewhere else.

Finally, we have also seen the power humans gain from reflecting on their own foresight. Once one realizes that anticipated scenarios are *just* representations, one is in a position to evaluate them, to modify them, to discount them, to discuss them, and to try to compensate for one's other various future shortcomings—including one's susceptibility to temptation.[59] There is as yet no evidence that nonhuman animals reflect on the limits of their predictions and prepare accordingly.[60] There are no signs of contingency planning—even on as simple a task as our forked tube. Nor do we see other telltale behaviors such as the setting of reminders to compensate for anticipated forgetfulness.

In modest memory tasks where we hide food under one of several cups and then wait before we give chimpanzees a choice, we do not see them simply marking the correct cup by, say, keeping a finger pointed at it. And while some animals act in ways that could be described as compensating for their future limits, these appear to be relatively fixed evolved systems. For instance, army ants leave pheromone trails and use them to find their way around, but will also keep running in circles if the trail accidentally gets into a loop—occasionally until they die.[61] By contrast, humans can use virtually anything to mark out routes and compensate for future forgetfulness, as well illustrated by Hänsel and Gretel dropping pebbles and bread crumbs to try to find their way home.

The serious limits we see in animal foresight may be the result of shortcomings in any one of the components from the mental theater or, perhaps more likely, in many of them.

———

The brute is an embodiment of present impulses . . . whereas a man's range of vision . . . extends far into the past and future.
—Arthur Schopenhauer (1851)

Were philosophers such as Seneca and Schopenhauer right when they made their sweeping claims about foresight being unique to humans? Assertions about human exceptionalism have a checkered history and may rightly raise suspicions about motives. Every species has characteristics that set it apart from other species. Many discussions about what makes us distinct appear less about science and more about justifying human superiority, divine origins, or the way nonhuman animals are treated.

When Darwin advanced his theory of evolution and confronted the world with the idea of common descent, the anatomist and paleontologist Sir Richard Owen, first director of the Natural History Museum in London (and famous for coining the term *dinosaur*), sternly defended the special status of our species. With the arrival in Britain of the first specimens of African apes, which Darwin believed to be our closest animal relatives, the public became increasingly curious about what set humans apart. Owen claimed that human uniqueness was evident in our distinct brain structures, which the apes were lacking. Thomas Huxley, who came to be known as "Darwin's bulldog" for his fierce defense of Darwin's ideas, subsequently demolished these claims, demonstrating that upon close examination, ape brains did in fact possess the same broad structural features as human brains. In an essay included in the second edition of Darwin's *The Descent of Man*, Huxley wrote: "There remains, then, no dispute as to the resemblance in fundamental characters, between the ape's brain and man's; nor any as to the wonderfully close similarity between the chimpanzee, orang and man, in even the details of the arrangement of the gyri and sulci of the cerebral hemispheres."[62]

Evidence of evolutionary continuity has not stopped philosophers, scientists, and laypeople from making new claims about what might ultimately set us apart. And frequently they entertain new names for the human species in that light. Here are but a few largely self-explanatory examples: *Homo aestheticus, Homo creator, Homo domesticus, Homo economicus, Homo generosus, Homo grammaticus, Homo imitans, Homo juridicus, Homo logicus, Homo metaphysicus, Homo reciprocans, Homo sentimentalis, Homo sociologicus,* and *Homo technologicus*.

Though *Homo prospectus* ("human who thinks ahead") has also been proposed, we are happy to stick with *Homo sapiens*.[63] And, in any case, many such suggestions have suffered the same fate as Owen's proposal about human uniqueness. For instance, Jane Goodall demonstrated that humans are not the only animals to use tools or to cooperate to kill members of their own kind.[64] Chimpanzees do both. Most claims that one trait or another sets humans apart have not borne out once the evidence was examined more carefully. Of course, it remains possible that future research will show that animals have greater foresight capacities than is currently evident.[65]

The animal kingdom does harbor an assortment of surprising abilities geared towards the future. In Queensland, Australia, there are jumping spiders from the genus *Portia* that appear to plan a hunt for other spiders, surprising their prey by rappelling down on top of them like a diamond thief in a heist movie. *Portia* spiders even avoid detection by taking detours or moving across the web of their prey only when wind disturbances offer a distraction.[66] There are undoubtedly many other apparently foresightful behaviors to be described in years to come in various other species. Think of how little we know about the cognitive capacities of, say, cephalopods—octopuses, squid, and cuttlefish—some of which appear to change color and behavior to communicate or deceive.[67] Like spiders, these creatures are invertebrates—they possess no backbone—and their capacities evolved quite independently from ours. More research is needed to figure out what they can do, and how they do it.

Humans share hundreds of millions of years of common ancestry with other living creatures. For this reason alone, one would expect humans to have many traits in common with at least our closer animal relatives. Darwin's concept of descent with modification implies gradual changes. Nonetheless, given that the last common ancestor that humans share with chimpanzees lived some six million years ago, it should also not surprise that distinctly human traits have emerged in many domains since that time.

Many species have evolved behavioral predispositions that equip them to act in preparation for recurring situations. Associative learning allows them to predict local regularities, at least if events are separated only by seconds. Recent evidence suggests that crows can even prepare for a specific future situation minutes in advance. But there is no compelling evidence that non-human species generate or communicate mental scenarios of remote future events, connect them into narratives, or reason about mutually exclusive versions of the future. Only in children's stories do animals conspire to rebel or plan to foil the bad guys' evil plots. Because only humans appear capable of such feats, our species has been able to dominate other animal life to our advantage—and perhaps to our peril. We are also uniquely faced with questions about how we should treat those not blessed (or cursed) with such foresight.

In the next chapter, we turn to what we know about when and how this capacity flourished in our ancestors. How did they discover the fourth dimension, steal fire from the heavens, and acquire the power to envision tomorrow?

CHAPTER 6

DISCOVERY OF THE
FOURTH DIMENSION

*Prometheus: "Still, listen to the miseries that beset
mankind—how they were witless before I made them have
sense and endowed them with reason. . . . Hear the sum
of the whole matter in the compass of one brief word—
every art possessed by man comes from Prometheus."*

—Aeschylus, between 525 BC and 456 BC

For the ancient Greeks, it was Prometheus, embodiment of foresight, who gave humans the godlike powers that distinguish us from other animals. Chief among these was control of fire. Archaeology tells us that our forebears learned to control fire much earlier than Aeschylus, author of the famed *Prometheus Bound,* could have possibly imagined—in fact, even long before those forebears became *Homo sapiens.* At one lakeside site in Israel called Gesher Benot Ya'aqov, archaeologists have excavated multiple levels of burned materials, including fruit, grain, flint, and wood, indicating that fire was used there repeatedly over the course of many years. The site has been dated to nearly eight hundred thousand years ago.[1]

Fire allowed ancient hominins to see when it was dark and to keep warm when it was cold. It gave them a formidable weapon to drive away predators and to smoke out prey. Once they had uncovered the secrets of fire, these people could destroy entire woodlands, radically changing the natural world around them.

Without foresight, nobody could have maintained a fireplace for long. Fire needs tending, after all, and combustible materials must be gathered in advance. At first, people would have had to keep an ember glowing continuously if they wanted to sit next to another fire tomorrow evening, for, once extinguished, it might have been a long wait before they came across another natural blaze. As illustrated in Jean-Jacques Annaud's award-winning 1981 film *Quest for Fire*, it would have been devastating for ancient hominins to lose the sacred flame they had long kept alive.[2] This dramatization follows three men who venture out to recapture the essential resource. After failing to secure a glowing ember from hostile groups, they rescue a woman from a tribe that has mastered the art of fire-craft. And so, they too learn this ultimate Promethean power.

Starting a fire is no easy feat—as anyone who has ever tried to get one going with sticks or flint can attest. Even when you know that banging certain rocks such as pyrites and flint together can create a spark, you must remain persistent and prepared to catch it. You need to have dry tinder at the ready to turn the flicker into a flame that can then be fanned. We do not know when exactly early humans discovered a way to ignite their own fire.[3] But we do know they eventually became farsighted enough to carry fire-making utensils with them in bags, just as the iceman Ötzi did, meaning they could look forward to the warmth of a campfire every night.

By spending time around fire, our ancestors also learned about many of its other useful effects. As we all know, various foods taste a lot better once they have been barbecued. Cooking with fire opened up new sources of nutrition, making roots digestible and ridding meats of parasites. In his book *Catching Fire*, anthropologist Richard Wrangham championed the case that cooking was a crucial evolutionary step.[4] Higher-quality food meant there was less need for constant foraging, freeing up precious time for other activities, while the caloric boost paved the way for the expansion of the brain. But fire not only affects edibility. Burning birch bark next to limestone cobbles, for example, can cause tar to build up on the stones.[5] This sticky substance can be used to glue stone heads onto wooden handles to make composite tools from more basic elements.

Huddling around a campfire every night set the scene for regular social exchanges and eventually sharing stories from the day. Telling tales allowed people to wire their minds together (satisfying our distinct urge to connect), to make sense of the past and better plot the future. And for people the world over, not just the ancient Greeks, legendary stories explain how their ancestors first became masters of the flame. The Wurundjeri people of the Kulin nation in Australia, for instance, relay the Dreamtime story of Crow, the trickster, stealing fire from the seven Karatgurk sisters—whose glowing firesticks can be seen in the night sky as the constellation Pleiades.[6]

To this day, our societies and civilizations are forged in fire, from heating our homes to propelling vehicles, and from generating electricity to smelting steel. But mastery of fire was just one of the dramatic transformative technologies brought about by foresight. Thinking ahead led to weapons and containers, maps and plows. Perhaps Aeschylus exaggerated when he had the fire-bringer Prometheus proclaim that he was the source of "every art." But this is probably not far off the mark, and in this chapter we will see that foresight must have been a prime mover in human evolution. As foresight evolved, new technologies and social dynamics emerged that, in turn, created pressures that themselves drove further advances in foresight.

———

The invention of tomorrow did not happen overnight. To understand how our ancestors came to think about the fourth dimension, we first need to consider gradual changes that occurred over vast stretches of time.[7] The big numbers needed to orient ourselves can quickly become confusing. Recall that the last common ancestor we share with modern chimpanzees lived some six million years ago. That is a mighty long time. But there are many other significant events that happened much earlier, as well as others that happened much more recently yet are still incredibly far in the past. It is tempting not to bother; to think of it all as one big, dateless *long ago*.

To help put things into perspective, imagine the four-billion-year his-tory of life on Earth condensed to, say, one month. On this compressed timeline, the first multicellular life forms appeared around a fortnight ago. And in the beginning of the last seven days, evolution created sex. Follow-ing the emergence of sexual reproduction, some animals eventually grew a spine about half a week ago. The first mammals appeared only yesterday, and the mammalian order we belong to, the primates, evolved about ten hours (or around sixty million years) ago. The vast span of time since we last shared a common ancestor with modern chimpanzees reduces to a more convenient sixty minutes at this scale.

A lot happened in that final hour to transform our ancient primate fore-bears into a species that mastered fire and, eventually, into people who can use fire to launch themselves on rockets into space. And we have not evolved in one straight line either; instead, our ancestral lineage is just one twig on a bushy family tree. Creatures on our branch of the tree of life are called *hominins*, which is the technical name for all human ancestors and cousins after the split from the line that led to modern chimpanzees. Figure 6.1 de-picts some of the most well-established distinctions.*

Our own species, *Homo sapiens*, first appeared on the planet only in the last couple of minutes on our condensed timeline.[8] Cave paintings appeared in the last thirty seconds, and the first solar calendars were devised in the final six seconds. The ancient Antikythera mechanism, the oldest computing device known, appears less than two seconds ago on this clock—and actual mechanical clocks were only invented less than half a second ago. Such re-cent achievements make us appear drastically different from other creatures, but for much of our prehistory we actually shared the planet with a motley crew of other bipedal hominins quite similar to us.

---

* Given the small fragments of bone usually found—and the detective-style nature of ar-chaeological research—there is uncertainty about exactly how many different species there were. We are here presenting distinctions that are currently well established. However, it seems safe to predict, given the frequency of recent archaeological finds, that more hominin species will be identified, and that genetic analysis will provide new clues about who is more closely related to whom on our family tree.

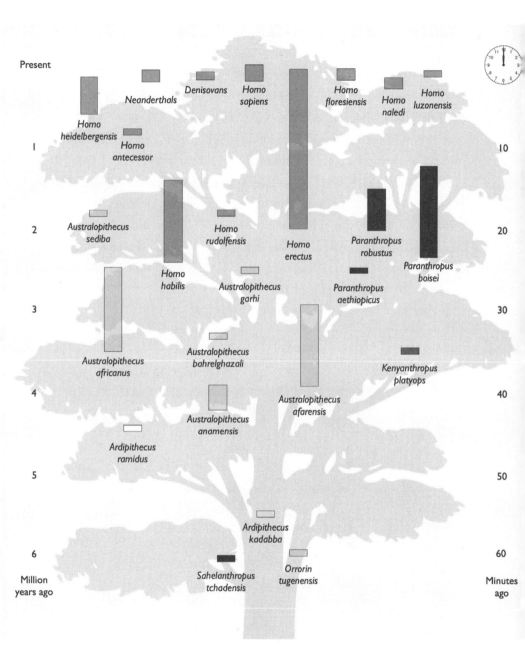

Figure 6.1. Hominin species of the last six million years (or sixty minutes on our condensed timeline). Long bars indicate that fossils from distinct points in time have been reported. Many of the shortest bars indicate that only fossils from one period (or even one specimen) have so far been found. We have not added any links between the bars because it is not certain which earlier species gave rise to which descendants.

Consider the situation some two million years ago (or twenty minutes ago on our timeline), when one of the earliest members of the genus *Homo* came onto the scene. *Homo erectus* already had a large brain with a cranial capacity about twice the size of chimpanzees'. Their body proportions were much like ours, and you would probably struggle to pick them out if you saw them walking in a modern crowd—though at a closer look, the sloping forehead and lack of a pointed chin might give them away.[9]

Homo erectus shared the African landscape with at least half a dozen other species in the hominin family, ranging from the small, stone-tool-wielding *Homo habilis* to the sturdy, strong-jawed *Paranthropus robustus*. Also present were *Australopithecus sediba*, the last of a genus of petite hominins that had long graced the African plains. Several of these species were so successful

Figure 6.2. Reconstruction of *Homo erectus*.

that they had survived for well over a million years. Perhaps it was to avoid competition from these upright-walking cousin species that *Homo erectus* left the African homeland and became the first hominin known to migrate throughout the Old World.[10] *Homo erectus* were pioneering forebears who left some tantalizing clues of early advances in foresight.

But let's start our journey back in time with a look at the common ancestor that lived an hour ago: the very last creature that could claim to be the great- (great-great-great-great-great-great-great- ...) grandparent of both humans and chimpanzees. Though we do not know much about this creature—the fossil record from this distant period is sparse—it must have been an ape adapted to life among the trees. What drove our forebears onto a different path?

It is commonly assumed that the critical first step was our ancestors leaving the forest, but instead it may have been the forest that departed. Around this time, tectonic plate movements began to bring about the Great Rift Valley that is still splitting East Africa from the rest of the continent.[11] The East no longer received enough rainfall to maintain rainforests, and so these apes were increasingly forced to move about on the ground, rather than swinging from branch to branch. As the predominant environment changed to the grassy plains of a savannah, our ancestors were confronted with radically different circumstances from their cousins on the other side of the rift.

Apes live quite a safe life in the trees, bar the threat of falling—and occasionally falling prey to climbing leopards. By contrast, the savannah is teeming with dangerous predators. To the large cats that roamed the grasslands, hominins would at first have seemed like an easy snack. But in spite of such monstrous threats, some of the apes stranded on the open plains evidently managed to survive.

How would you deal with being on the same grassland as lions? Say you could take one tool with you, what would it be? Many people would no doubt pick a loaded gun, *thank you very much*. What such a weapon lets you do, of course, is hurt the predator before it is upon you. Perhaps early hominins had a similar idea. Obviously, they had no firearms, but they could have thrown objects at a threat. In addition to hiding or climbing, hurting at a distance could have offered a new way out of trouble.*[12] Apes can hurl objects in fits of display, as the chimpanzee Santino did, though they do not have particularly good aim—a bit like a discus thrower trying to hit a bull's-eye. One hominin stone thrower with poor aim would have had little chance of driving away a hungry lion. However, a group of hominins doing this in unison might have given the cat pause to think twice about pursuing such an annoying (and skimpy) prey animal. That antelope across the plains might have seemed like less of a hassle.

Initially, our forebears may have chucked whatever was available, such as sticks and rocks, to fend off predators. But if nothing useful was lying around, this defensive strategy was not an option. Selection pressures might have therefore favored the prepared.[13] Individuals who had a tendency to carry rocks would have had a higher chance of surviving attacks than those that hurriedly sought a suitable projectile only once a predator was bearing down on them. Similarly, those that traveled in groups and defended themselves cooperatively would have had a selective advantage over more solitary individuals.

So the first steps towards increasing preparation and cooperation might be traced to how our ancestors adapted to live in a savannah environment. Those that avoided becoming dinner might have been more foresighted and collaborative than their companions, detecting threats earlier and devising more effective defenses in advance. Or, to put it the other way around, the sharp claws of natural selection may have weeded out the unprepared and uncooperative. No matter how one puts it, this is only a plausible account—a just-so story—not one that has unequivocal archaeological support. Though analyses of hand anatomy support the idea of early adaptations for throwing (and clubbing), exactly when hominins first carried projectiles, even just stones to throw, has not been ascertained.[14]

Whether or not early hominins were carrying rocks, evidence from Lomekwi in Kenya shows that by 3.3 million years ago they were using

---

* Hurting at a distance is unusual but not entirely unique to humans. There are various animals, from archer fish to spitting cobras, that can do this for attack or defense. The velvet worm *Onychophora* can squirt glue-like slime to catch prey and may have done so for millions of years.

rocks to hit other rocks.[15] Whereas chimpanzees use stones to crack open nuts, these hominins used a similar technique to make stone tools. What exactly these early tools were used for is still unclear. Other small stone tools—Oldowan tools—appear in the archaeological record from 2.5 million years ago, and cut marks on bones indicate that these tools were used to butcher carcasses.[16] Oldowan tools are associated with *Homo habilis*, the first members of the genus *Homo*. It is likely that these early humans—species in our own genus we will also refer to as *archaic* people—made many other tools out of more perishable materials, like wood, that we have no trace of today. But, as yet, there is relatively little evidence to suggest they prepared any tools for use in the more distant future.[17]

This changes with the emergence of Acheulean tools associated with *Homo erectus* some 1.8 million years ago. These tools, named for their initial discovery at Saint-Acheul, in France, include bifacial handaxes which are symmetrical in structure. Making these tools requires a multistep plan. From selecting the right raw material to delivering the precise blows that make the object so well balanced, the toolmakers needed to think ahead—to envisage what they were trying to achieve before making it a reality. The creator of an Acheulean tool would first wield a hammerstone to fashion the rough shape of the end product before working around the circumference of the core, removing flakes with strikes that alternated between each side. Next, the tool was thinned with careful blows. Finally, the craftsperson needed to trim and shape the regular edge and symmetrical form. If you ever have the opportunity to try it yourself, you will see how sophisticated a skill this really is. Recall that in one of the studies we encountered in Chapter 2, modern students of this technique still did not fully master it even after dozens of hours of training.[18] Put simply, these tools indicate a sophisticated degree of foresight unlike anything seen in other animals.[19]

Because it takes time, effort, and skill to make Acheulean tools, it is perhaps unsurprising that they even appear to have been carried over distances and used repeatedly.[20] This suggests considerable premeditation. However, at Olorgesailie in Kenya, a site excavated by the famous paleoanthropologists Louis and Mary Leakey, the ground is still littered with thousands of these

Figure 6.3. Bifacial handaxes are part of the Acheulean tool kit associated with *Homo erectus* and *Homo heidelbergensis*. These tools required considerable foresight to make and were transported and used repeatedly.

primordial teardrop-shaped stone tools. Why did *Homo erectus* apparently abandon so many perfectly good ones?[21] If they needed a tool, they could have just picked one up. One possibility is that they were practicing their manufacturing skills.[22] Once they were proficient, these early people could have wandered the plains knowing they could make a new tool if their old one broke. *Homo erectus* not only had all the right anatomy to throw, hit, and run—they may well have been armed and ready to reload.[23]

For well over a million years—some three hundred times as long as all of recorded history—ancient hominins busily went about crafting these versatile bifacial stone handaxes. *Homo erectus* were thinking ahead in a new way and were evidently capable of learning from one another, passing on knowledge from one generation to the next. But this period is also marked by a distinct lack of innovation. Remarkably little about the construction of these tools changed over thousands of generations, except that the handaxes became very gradually slimmer and more refined.

It is hard to imagine modern humans, with their penchant for innovation, hanging on to the same tradition for even a tiny fraction of that time. Someone would attach a stick to the handaxe or hammer it into a different shape—and these techniques, if they made a better tool, would quickly catch on. Pick any dominant weapon of more recent eras and you will typically find it was refined or replaced within a few generations. Despite the evidence that *Homo erectus* made significant advances, their mental time machine and culture appear to have been quite distinct from ours. The dynamic feedback loop between foresight and culture had not yet gotten into full swing.

———

In 2018, further studies at Olorgesailie documented that a dramatic shift in stone tool technology finally occurred sometime before three hundred thousand years ago. The tools were now smaller, including more precise blades and sharper points. The material used to make tools also changed, with many crafted from black obsidian, a shiny volcanic glass that originated over fifteen miles away. In fact, one piece appears to have been sourced from a location sixty miles from its final resting place: more than the length of two marathons. This suggests considerable commitment to transporting the material (or an early network of trading).[24]

This is the time when archaic humans known as *Homo heidelbergensis* were prevalent. First documented near Heidelberg in Germany (hence the name), these people stood some six feet tall, had thick brow ridges, and had a skull large enough to house a brain at the lower end of the modern human range. They lived from around 600,000 to 150,000 years ago, with fossils found in Africa, Asia, and throughout Europe.[25]

Homo heidelbergensis began to use a new, even more farsighted technique for making stone tools. Unlike Acheulean tools, these Levallois tools were made by creating sharp and flat flakes from a core.[26] The production required a longer series of complex preparatory steps. First the toolmaker would carefully shape the stone core and create a striking platform, and only then deliver a forceful detachment blow that separated a target flake, which could then be used, for instance, as a skinning knife. By transporting

Figure 6.4. Reconstruction of
Homo heidelbergensis.

a shaped core, rather than a single handaxe, archaic people were prepared to make further flake tools when they needed them.

With so many people striking so many different types of rocks together over so many hundreds of thousands of years, it might not seem surprising that eventually someone noticed sparks flying. It is around this time that archaic people began to use fire more frequently, and evidence for heat-treated glues and composite tools appears.[27] The innovation of composite tools led to endless new possibilities. Making tools no longer involved merely removing something from an object, such as knapping a stone flake off a core, but also adding elements together into novel arrangements, such as attaching a stone point to a spear.[28]

These advances suggest changes in the minds of our ancestors. A find of three-hundred-thousand-year-old spears near Schöningen in Germany, together with the remains of butchered horses, demonstrates that *Homo heidelbergensis* could form hunting plans that enabled them to bring down mighty prey.[29] Although it was not at first clear whether these spears were thrown or simply thrust, modern javelin athletes testing a replica found they could hit a target at a distance of over sixty-five feet.[30] A report in 2020 documents signs of repeated high-pressure impact on a carved stick from the same site—strongly suggesting it was thrown and not just used as a club or handheld weapon.[31] Hurting at a distance had taken off.

At Atapuerca in Spain, there are indications that archaic people from the same period may have discovered a new hunting technique that avoided combat altogether. At a cave site called Galeria, a roof collapsed and created a shaft more than thirty feet deep.[32] Bison and other animals repeatedly fell into the hole and were subsequently defleshed by the cave dwellers. These may have at first been just happy accidents of meat falling from the ceiling, but they might have eventually given people an idea for a plan: Why not simply spook the prey into running towards a sudden vertical drop? By preparing traps like these, hunters would have been safer and no longer limited to targeting the old, young, and weak of a herd.[33]

Taken together, the evidence suggests that from at least three hundred thousand years ago onwards, then, major changes were afoot. With the appearance of Levallois tools, archaic people finally moved beyond the relative stagnation of Acheulean technology that they had relied upon for over a million years beforehand. Archaic humans became top predators, not with brute force alone but also with a mind capable of more potent foresight and the innovation and coordination this facilitates—features that in turn enable even more control over the future.[34] Evidence that these people had begun to transport materials over long distances, craft composite tools, and hunt big game with complex strategies suggests that the feedback loop between foresight and cultural evolution was beginning to gather momentum.[35] The scene was set for another innovation that opened the door to a whole new niche for our ancestors to exploit.

———

"The more toys you bring back, the more stickers you will get from Tigger." This was our simple instruction when we presented three-to-seven-year-old children with a table full of toys and asked them to carry as many as they could to another room.[36] Older children readily made use of the woven basket we had placed in plain sight among the objects. Nearly all the three-year-olds, on the other hand, failed to use this carrying device and ended up transporting only the few toys they could bundle up in their arms. The power of carrying devices is easily overlooked, not just by young children but also by scientists. Yes, it is widely acknowledged that some kind of bag or basket would be useful when gathering foods, but these tools undoubtedly did much more for our ancestors.

Carrying devices are an exemplar of the foresight-culture feedback loop in action: not just a reflection of increasing planning capacities but themselves facilitators of even more potent preparation. We learned in Chapter 2 that the key to innovation is recognition of future utility: foreseeing that something will continue to be, or might turn out to be, useful in the future. Investing time and effort into crafting ever more sophisticated tools may not pay dividends when each tool is used only once. By contrast, if you can retain tools, you can use them again and again—in fact, you have access to them even when the raw materials are nowhere to be seen. But the more objects one possesses, the more apparent become the limits of having only two hands for carrying them. With the invention of carrying technologies—slings, sacks, and other containers—people could offload the task of holding on to materials for longer intervals and over farther journeys.

Today, we all depend on carriers to transport objects that may potentially become useful at some future point. They are everywhere: from the trillion plastic bags used each year to million-dollar designer handbags. We regularly embed containers within other containers, such as when we put pillboxes into toiletry bags, toiletry bags into suitcases, and suitcases onto trolleys. Anthropological data suggest human groups across the globe have long relied on strings, nets, straps, and many other techniques for increasing their carrying capacity. The ingenious solutions range from the gourds used by the Hadza in Africa to store honey to the maize husk baskets woven by many Native American peoples, not to mention the baby slings used in nearly every human culture.[37]

The initial idea to use a device to transport some other object would have been a critical insight to get the accumulation of material culture up and running. After all, mobile containers are metatools—tools that can serve as tools for other tools. With the invention of carrying devices, people could keep entire tool kits on their person. From then on, humans could make sure their local habitat included whatever was needed wherever they went: a portable life-support system, as it were.

Having tools and weapons at the ready would have been critical in plenty of situations. The extensive effort required to make a hafted handaxe could now pay off again and again. With the innovation of carrying devices, there was serious benefit to be gained from dedicating time towards making many tools—axes, knives, or spears—in advance. When humans encountered a source of excellent raw materials, say some smooth black volcanic glass ideal for making sharp arrowheads, it now made sense to bag that material or to sit down and start knapping right away. Even implements that would only find an occasional use became worthwhile to prepare and bring along—just in case.

Mobile containers created a new niche that drove humans to invent ever more tools. Recall that Ötzi carried dozens of tools in his quiver, basket, and backpack: without them, it would have been nigh on impossible for him to survive for very long in the harsh climes of the upper Alps. Without this seemingly simple innovation, and the diversification in human tool-use it enabled, people would have been unlikely to flourish in the wide range of extreme and dangerous environments we now inhabit. As we have seen, humans are generalists who can compete with narrowly evolved specialist species—with adaptations finely tuned to the particulars of any environment—simply by carrying specialist equipment and knowing how to use it. So significant was this advance that it might be tempting to rename us *Homo baggins*—but let's not get carried away.

Other animals have natural containers, such as marsupials who carry their offspring in their pouches or pelicans who carry fish in their throat sacks. And, of course, humans can get animals to carry things, putting a satchel on a donkey, a message on a pigeon, or a barrel of brandy around the collar of a Saint Bernard. There is no doubt that other species have the physical capacity to carry mobile containers, but there is no good evidence that they have the foresight required to recognize the future utility of such devices.

Together with archaeologist Michelle Langley, we recently reviewed the available evidence for the emergence of mobile containers in human evolution.[38] The two earliest known examples of hominins using containers—shells to store pigments—date from around one hundred thousand years ago; one is from *Homo sapiens* in South Africa and the other is from Neanderthals in Israel. Signs of mobile containers made from wood appear by about fifty thousand years ago. Of course, baskets and netted bags decompose rapidly, so there is little chance of traces surviving for many millennia. It was exciting, therefore, when in 2020 a study of a Neanderthal site documented even a small, scraggly piece of three-ply cord made from bark fibers dated from about the same time.[39]

Eventually, some modern humans left Africa and the Middle East and migrated across the Old World—with or without backpacks.[40] They reached Sahul (ancient Australia and New Guinea) soon after. The peopling of such landmasses that were separated by large swathes of ocean suggests the use of watercraft. And boats, in a sense, are mobile containers for carrying humans and other cargo. The oldest signs of archaic humans occupying islands, in the form of stone tools and fossils from Indonesia, reach back hundreds of thousands of years.[41] However, it is of course possible these people merely clung on to debris trying not to drown, before finding themselves washed up on new shores.

Compelling evidence of planned watercraft usage comes from the transport of obsidians from the volcanic island of Kozushima to the Japanese mainland.[42] Some thirty-eight thousand years ago, modern humans brought back large amounts of the volcanic glass to make tools. The island was never joined to the mainland, and even with a massive drop in sea levels, it would still have been at least twenty miles away. This finding makes it clear that people were deliberately transporting material by traveling both ways across challenging waters.[43] These journeys would have taken many hours and required considerable skill, technology, and planning.

With the innovation of mobile containers, humans not only kickstarted drastic changes in material culture but also carried themselves into new worlds of possibility.

———

When did our ancestors finally evolve modern minds, equipped with mental time machines like ours? In an influential review, paleoanthropologists Sally McBrearty and Alison Brooks identified the following four characteristics as essential to "modern behavior": sophisticated planning, innovativeness, abstract thinking, and the use of symbols.[44]

This analysis affirms the important role of foresight in modern behavior. Without foresight there could obviously be no sophisticated planning, and as we saw in Chapter 2, our solutions to problems would not be innovations if we could not recognize their future utility. Even abstract thinking and the use of symbols are closely linked to foresight. We saw in Chapter 4 that we can use abstract concepts as elements to combine when constructing scenarios of the future (and of course we can think about *tomorrow* without picturing anything concrete). The symbols we use the most are words, and linguists suggest a defining feature of human language is the capacity for *displacement*—to refer to things that are remote in space and time.[45] Symbols are *about* something other than themselves, and we often use them to inform others about the future. Road signs tell you what to do. Flags tell you who you can find here. And maps tell you what you can find where.

Signs of modern behavior appear en masse in the European archaeological record from about forty thousand years ago onwards.[46] Cave paintings such as the beautiful rock art in the Chauvet-Pont d'Arc cave in France, sculptures such as the finely carved Venus of Willendorf figurine discovered in Austria, innovative new technologies like bone tools and microlithic blades, and even musical instruments such as flutes make a first appearance.*[47] Together, all of this ostensibly pointed to a radical change: the beginning of "modernity."

Yet anatomically modern humans, featuring our characteristically high and rounded skull, a bony chin, and small brow ridges, had existed for well over a hundred thousand years by that point. The relatively sudden, late

---

* In the caves of Isturitz in the French Basque country, the sound of some of the oldest flutes eerily reverberating in the cathedral-like chamber creates a magical atmosphere that may have been further enhanced by drumming on the stalagmites, and the cave paintings and etchings of animals flickering in the light of bonfire. Our ancestors created new worlds of sound and vision.

emergence of "modernity" was therefore perplexing. Was this apparent revolution the result of a fortunate genetic mutation? Was it the outcome of the
discovery of language? Or was it the product of some other radical advance
in cultural evolution?

It turns out none of these widely debated answers is correct. In their seminal paper, McBrearty and Brooks demonstrated that there was no sudden
revolution. Yes, there is a wealth of new behavior evident in the European archaeological record from this period, but this likely reflects the arrival of modern humans on that continent. When archaeologists looked more carefully
outside Europe, they discovered signs of a gradual rise of behavioral modernity
in Africa and Asia starting much earlier—and we now know the first Australians maintained a continuous culture that has lasted sixty-five thousand
years or longer.[48] A groundbreaking excavation of the Madjedbebe rock shelter
in Australia's Northern Territory led by archaeologist Chris Clarkson found
grinding stones, ochers, and hatchets dated to at least that old.[49]

Among the earliest signs of modernity outside Europe are sophisticated
new tools, including barbed bone points and a bone knife, used some
ninety thousand years ago in what is today the Democratic Republic of the
Congo.[50] The oldest evidence of ceremonial burials, and hence perhaps of
planning all the way into an afterlife, is even slightly older.[51] Other ancient
indications of modernity were found at the Blombos cave in South Africa.
Pieces of ocher pigment engraved with abstract geometric designs, including
crosshatched patterns, indicate symbol use over seventy thousand years ago.
It has even been suggested that these patterns may represent instructions for
weaving techniques. Other signs of modern behavior include bone tools as
well as shell beads that were deliberately perforated, presumably to attach
them to a thread. Such decorations are a first sign of the venture to shape future physical appearance as we discussed in Chapter 3.[52] One study suggests
shells found in Israel were suspended on string as bodily adornment perhaps
as early as 120,000 years ago.[53] Recent findings have thus confirmed that
many signs of modern behavior appear long before *Homo sapiens* reached
Europe.

The oldest known human representational art has been found on the
Indonesian islands of Sulawesi and Borneo. There, the archaeologists

Maxime Aubert, Adam Brumm, and their colleagues discovered a forty-four-thousand-year-old cave painting of wild pigs and buffalo being hunted by human-animal morphs sporting tails and snouts.[54] This is also currently the oldest compelling sign of supernatural thinking, and indeed of any kind of depicted narrative. Similarly, Michelle Langley and her team have reported evidence of forty-eight-thousand-year-old portable art (including shell beads, perforated ocher nodules, and silver pigments) from Sri Lanka, alongside innovative bow and arrow technology.[55] Together, such discoveries have painted a picture of much more gradual, dispersed, and ancient origins of behavioral modernity—and the foresight it entails.[*56] The hypothesis of a sudden human revolution in Europe has been comprehensively falsified.

———

In the years after the naturalists Charles Darwin and Alfred Russel Wallace independently discovered evolution by natural selection—Darwin having been compelled to publish upon learning Wallace was also onto the idea—both men reflected on the implications of their new theory for the origins of humankind. While Darwin initially avoided openly discussing human evolution for fear it would be too controversial, Wallace tackled the issue directly. Despite natural selection explaining the diversity of life on Earth, Wallace eventually concluded that the human mind must have nonetheless been bestowed by God or another superior intelligence unbound by the shackles of known natural laws. That's because, Wallace reasoned, the extraordinary powers of the human mind, including "those faculties which enable us to transcend time and space," are so much more developed than necessary to merely survive and reproduce in our ancestral environment.[57] After much handwringing, Darwin eventually disagreed, famously responding: "I hope you have not murdered too completely your own & my child."

---

* What's more, these signs of behavioral modernity may not have even been unique to *Homo sapiens*. Neanderthals also sometimes buried their dead. And some surprising new finds have challenged the idea that it was modern humans who first brought symbolic behavior to Europe. For instance, the oldest sign of bodily adornment in the world now comes from Neanderthals in present-day Croatia, where eagle claws appear to have been mounted into a necklace some 130,000 years ago. "Modernity" was therefore not unique to *Homo sapiens*— at least as indicated by such symbolic behavior.

In *The Descent of Man*, Darwin finally addressed the issue of human origins head-on. When he forewarned Wallace about the book, the latter replied: "I look forward with fear & trembling to being crushed under a mountain of facts!"[58]

Today, the facts of human evolution are utterly overwhelming to anyone willing to fairly examine them. And, in spite of many question marks, there are also entirely plausible accounts of how our mental capacities, including those that enable us to "transcend time and space," evolved.

To account for any lingering concerns about how evolutionary processes could have brought about such exceptional minds, we want to draw your attention to two key points. First, adaptations can often themselves create new sets of pressures that drive the selection of further adaptations. Second, apparently vast gaps between traits in currently living species can be the result of intermediate forms having gone extinct.

On the first point, let's return to the prehistory of fire, before we deal with the second point below. Control of fire changed the world of early humans and their descendants—just as beavers' dams or termites' mounds create new environments with new selective pressures for their descendants.[59] Once hominins had some modicum of control over fire, this would have created new advantages for those who could exploit this resource in more effective ways. Individuals with the requisite foresight could have reaped benefits by deploying fire in hunting, defending, cooking, and many other contexts—meaning whatever mutations increased a propensity or ability to do so could now be selected for. Thus, control of fire, itself a product of enhanced foresight, might have introduced new selection pressures for bigger brains capable of yet more advanced foresight.

Once people began congregating around the fire each night, this would have also created social forces that did not exist when everyone retired with the sun. Imagine what early people might have thought while staring into the flame, night after night, generation after generation. If they had encountered something important during the day, like a dangerous predator or a herd of prey animals up ahead, recollections may have bubbled up as their eyes were transfixed by the flickering light. And this may have sparked glimmers of reflection and planning. Clearly, great benefits could be reaped if one could somehow relay such information to the others in the round, so that tomorrow the threat might be avoided or the opportunity seized.[60]

The psychologist Merlin Donald has proposed that the first steps in the evolution of more effective communication and cooperation might have involved deliberately reenacting events.[61] Barking like a dog or bounding like an antelope may have initially sufficed to get the basic idea across, especially when also pointing in the relevant direction. And with the first successes, there would have been new selective pressure to make this communication ever more efficient. Perhaps the pantomime required to get across the idea of, say, a lion could have conventionalized to a quick roar and the raising of a "paw," as gradually representations became ever more abstract. What is at stake here, then, is really the evolution of language itself. Speech may well have had its roots in gestural communication, as Michael Corballis long argued.[62] Even though this itself is another just-so story, it should suffice to illustrate how one new behavior can beget many other downstream changes. Eventually, and to this day, people sit by the fire recounting what happened during the sunlight hours, spinning yarns, and plotting tomorrow.

The comparative and developmental psychologist Michael Tomasello, having long researched the capacities of great apes and young children, has concluded that what makes humans unique is the way we link our minds together and cooperate—what we called the *urge to connect* in the previous chapter.[63] He highlights how humans reason so expertly about the minds of others, how they communicate, and how they ask questions and give advice, as well as their fundamental desire to help one another. However, it is often overlooked that without foresight, much of human cooperation would be impossible. Promises, for instance, require people to bind themselves to particular future actions. Agreed goals, collective teaching, obligations, and many other aspects of human interaction depend on our virtual time machines.

One of the earliest signs of advanced communication and coordination operating in tandem with foresight comes from *Homo erectus*. At Gesher Benot Ya'aqov, archaeologists not only documented that these hominins controlled fire, they also found evidence that they butchered an elephant with stone axes. The odd thing is that these people somehow appear to have also managed to overturn the elephant's massive skull to gain access to its brains.[64] This was probably not a one-person job. A boulder and a log found underneath the skull even raise the possibility they used a lever to accomplish the task. However they did it, the feat probably required the coordinated efforts of several individuals pursuing the common goal of picking up and rotating the head.

The archaeologist Ceri Shipton and the psychologist Mark Nielsen have argued that a 1.2-million-year-old site in India even indicates division of labor in *Homo erectus*.[65] Here, single stone tools were made at two distinct clusters at the same site: one location where large flakes were knocked loose and another where they were retouched into their finished forms. This makes little sense if one individual was making the tools, but it makes a lot of sense if the separate tasks were performed by different individuals.

There are other hints that *Homo erectus* individuals may have helped each other out and coordinated their actions. A nearly toothless skull from Dmanisi in Georgia featured advanced bone loss after the teeth had fallen out.[66] This suggests the individual somehow managed to survive for some time without being able to chew—perhaps others did it instead. More signs of such social support become evident in other hominins as we move forward in time. For instance, a five-hundred-thousand-year-old *Homo heidelbergensis* spine shows a hunched back.[67] Despite considerable incapacitation, this individual lived to a ripe age, again suggesting others helped. By the time of the Levallois technology, as we saw, there is increasing evidence of cooperative abilities, such as collaboratively hunting horses with spears.

Living in a group that relies on collaboration and foresight creates further pressure on becoming even more collaborative and prudent. In a sense, as some scholars put it, humans effectively domesticated themselves to become more tolerant and helpful. When modern humans domesticate other animals, such as dogs, they tend to breed for friendly and nonaggressive traits. This *artificial selection* also tends to result in various physical changes, such as smaller skulls and teeth, alongside more general compliance. Ancient hominins may have done something similar to themselves. Relying on cooperative defense, hunting, and eventually sharing a place around the fire, they may have increasingly acted to reward collaboration and punish transgressions—and they may have found people with a more friendly disposition more attractive. In turn, individuals who had the foresight to realize that selfish or antisocial behaviors may not pay off in the long run would have been more likely to collaborate and be liked by others.[68] As Richard Wrangham put it: "There was active selection, for the very first time, against the bullies and the genes that favored their aggression."[69]

This is not to say that bullying and aggression were rooted out entirely, of course. In fact, the darker side of our nature may have been a key force in our story. Yes, humans are hypercooperative and helpful, but we can also be incredibly effective aggressors, and we can even engage in friendly collaboration in pursuit of such violent goals as the complete annihilation of a common enemy. Though perhaps less pleasing to ponder, it is likely that conflict and violence were also important factors influencing the survival and reproductive chances of our ancestors—in the process driving the evolution of foresight and culture.

———

Neither Wallace nor Darwin was aware of other hominins. But we now know our ancestors were hardly ever alone, instead sharing the planet with a diverse crowd of upright-walking cousins. What happened to *Paranthropus* and *Australopithecus*? These smart, tool-using species had been successful for very long periods of time. And what happened more recently to Neanderthals, who seemingly shared some of our key modern traits? The extinction of all other hominins has left a vast apparent gap between humans and the rest of the animal kingdom. This is the second point to consider when asking why we now appear to be so exceptional.

In light of the big picture of the evolution of life on Earth, the demise of this multitude of hominin species is not particularly unusual. It is widely estimated that some 99 percent of species that have ever lived have gone extinct.[70] Though extinctions happen for many reasons, and each hominin extinction probably had diverse causes, such as diseases, dwindling resources, and changing climate, it is possible there was another distinct factor at play. Our own ancestors may have had a hand in our cousins' demise. Competition and even outright violent conflict may have made the difference. Those with better innovations, more effective coordination, and shrewder plans may have won the day, again and again.

It is uncontroversial that humans are currently causing the extinction of many species—even if most humans probably consider this highly objectionable. There is widespread recognition of an extinction crisis.[71] Though likely not with the same speed, human activity may also have been a culprit in the extinction of many other animal species in the past tens of thousands of years.[72]

In Chapter 2, we noted it is probably no coincidence that the arrival of humans in new habitats was frequently followed by the mass extinction of large animals. In his book *The Future Eaters*, the paleontologist Tim Flannery vividly outlines how migrating humans brought about such changes.[73] The last major habitable landmass to have been discovered by humans was New Zealand. When the Maori first arrived on canoes some seven hundred years ago, they found a land of plenty.[74] Though they had brought chickens with them, they stopped breeding them once they discovered the large flightless poultry foraging in the untouched forests. These *moa* had little chance against the new threat, being unaccustomed to contending with mammalian predators. With the destructive energy of fire and the solid force of stone tools, the Maori quickly and dramatically altered the environment. The moa hunters flourished . . . and the birds went extinct, perhaps in as little as two hundred years.

Whereas New Zealand had been uninhabited when the Maori first stepped onto its shores, many of our migrating forebears would have ventured into lands already occupied by other hominins. Recall that *Homo erectus* had left Africa well over one and a half million years ago. When modern humans finally left Africa and the Middle East only some seventy thousand years ago, they would have encountered the descendants of earlier migrations, including the small-statured "hobbits" of Flores in Indonesia, the newly described *Homo luzonensis* in the Philippines, the Denisovans in Central Asia, the Neanderthals in Europe, and possibly a range of other hominins whose remnants have yet to be discovered.[75] In all those cases, without a virgin land to exploit, the newcomers had to come to an arrangement with the locals or compete for limited resources. This could have led to direct conflict. As foresight enabled our ancestors to increasingly cope with the old predation challenges from bears, wolves, and tigers, other hominin groups with similarly developed capacities to think ahead eventually became the most dangerous threat.

Here we see again how adaptations create new sets of pressures that in turn drive the selection of further adaptations. Foresight gave our ancestors new reasons to fight and led to new arms races, even quite literally. The seventeenth-century philosopher Thomas Hobbes proposed that the reasons for our quarrels can largely be subsumed into three fundamental categories.[76] The first category comprises the obvious grounds of conflict over some kind of limited resource, such as food, water, territory, or mates. Unsurprisingly, aggression for these reasons is not uniquely human. Humans may merely take this to extremes when we cooperate to attack others to take what we want.

As we noted earlier, chimpanzees sometimes cooperate to kill members of their own kind, suggesting our violent roots go very deep. Jane Goodall reported that one murderous pair of chimpanzees attacked several mothers in their own group.[77] On one occasion, the pair viciously wounded a new mother, and eventually killed and consumed her infant. Despite the ongoing cannibalism, within fifteen minutes the still-bleeding mother approached the pair and offered a hand in reconciliation. This is not what you would expect a human parent to do at this point. What about retaliation? Retribution? Revenge? Reprisal? As our collaborator social psychologist Bill von Hippel has remarked, human parents in such a situation would look at each other knowingly and conspire to put an end to the menace.[78] But no, none of the victims did any of this. They made up and moved on. Alas, the attacks from the marauding pair continued over a period of three years. Why did the victims not gang up and bludgeon the culprits to death while they were asleep—before they could commit another attack?

Human foresight leads to aggression in situations where chimpanzees do nothing. Hobbes's second category of violence arises from the capacity to think ahead and conclude that you or your group members might become victims of aggression in the future. *That pair may well try to kill my next baby.* Contemplation of future possibilities drives us to prepare for potential assault by building defensive structures or sharpening weapons. What's more, it can lead us to take the initiative and strike first. Preemptive strikes make sure the anticipated conflict happens when and where the odds are favorable—and often in conspiracy with others, such as previous victims and potential future victims.[79] Mutual distrust can drive people to build up arsenals and attack each other even when neither party actually wanted to pursue aggression in the first place—a so-called Hobbesian trap. Once early humans were pursuing proactive violence, natural selection would have favored individuals better able to anticipate others' plans and to consider what others might expect about their own plans. Violent redress can of course breed yet further violence, especially when there are two groups that both remember and think ahead. This can account for the escalating and protracted blood feuds that litter human history.

At first sight, Hobbes's third category of reasons for aggression appears to be about trivial matters. People often fight over cheating in games or cutting a line, over rude gestures and other signs of disrespect. The underlying driver for such conflicts, as Steven Pinker points out, flows from the other two reasons for violence. The best way to deter someone from an attack, preemptive or otherwise, is to demonstrate a willingness and capacity to retaliate. The key for this to work, however, is that the other party can think ahead too and consider the possibility of later retaliation to be a serious threat. Credible deterrence is built on reputation. People therefore often settle scores in public so as not to undermine their honor and display signs of weakness. When there are no witnesses, it is much easier to let go or back down. Credible deterrence is a key factor in many international stand-offs—just think of the Cold War threats of mutually assured nuclear destruction.[80]

In addition to giving us more reasons to fight, foresight made fighting more dependent on reason. From history it is clear that success at fighting, however it got started, is often not only a matter of strength and skill but also of wit. Strategically planned approaches to conflict tend to be victorious over charging in without organization or a plan. Accordingly, coordinated fighting forces came to dominate much of our recorded past, from small marauding bands raiding and plundering to large standing armies invading or defending empires. Remnants of fortifications—walls, moats, castles, and bunkers—are solid reminders that humans have long realized other humans are a threat worth preparing for.

The record of human warfare is full of plots, trickery, and deceit. Conflicts were often won by people who excelled not only at foresight and cooperation per se but also at thinking about their own and others' foresight. Even mighty, meticulously organized forces have at times been held back or defeated by weaker but more cleverly planned opponent strategies. From feigning retreat to leaving a wooden horse at the gates, cunning deception can be used to set up an advantage the enemy could not foresee or to make them falsely predict what one plans to do. Analyses of modern conflicts in small-scale societies suggest that the most common form of attack is a planned ambush, and in such fights the casualties tend to be overwhelmingly on the defending side.[81] As Sun Tzu, in his fifth-century-BC treatise *The Art of War*, proclaimed: "Attack him where he is unprepared, appear where you are not expected. . . . The general who wins a battle makes many calculations in his temple ere the battle is fought."[82]

We need to be careful when generalizing from history to prehistory, of course. But it should be clear that the emergence of enhanced foresight, and the new forms of cooperation and competition this enables, can, in turn, create new forces that drive the selection of yet further enhancements in foresight. This is one of the reasons why human abilities may appear so much more sophisticated than they need to be in order to "merely" survive and reproduce.

The oldest known indication of a hominin being killed violently by another is from over four hundred thousand years ago. A skull from Atapuerca in Spain was found with two very similar impact fractures on the frontal bone. A hole in the head can happen in many ways, such as from a fall onto a rock, but two virtually identical holes suggest this was no accident: someone appears to have hit the person with the same solid object twice in rapid succession.[83] The extent of violent conflict between our ancestors and other hominins is difficult to discern. There are only a few tentative indications, such as the cut marks on a jaw of a Neanderthal child in a cave in southwestern France that imply slaughter—while the other remains in that cave are all from modern humans. Though suggestive of gruesome interspecies violence, such findings are by no means conclusive. The picture of modern humans fighting modern humans is much clearer, at least since the end of the last ice age.[84]

Surely not all encounters between our ancestors and other hominins would have been violent. At times, foresight on two sides may have allowed not just peaceful coexistence but prosperous relationships. It is easy to see how capacities for planning, innovating, teaching, preparing, and strategizing could have also helped them carve out a mutually beneficial arrangement with the neighbors. Whether peacefully or not, Neanderthals and modern humans may at times have lived side by side, for instance, in caves at the Mediterranean coast south of today's city of Haifa.[85]

Genetic evidence even shows that Neanderthals and modern humans had offspring. If you are of European or Asian descent, you are most likely also of Neanderthal descent. In a sense, Neanderthals did not entirely die out, because a fraction of their DNA lives on.[86] What's more, modern humans also interbred with other hominin species. If you have Melanesian heritage, you may have also inherited some DNA from Denisovans.[87] It seems only a matter of time until the remains of yet other ancient ancestors will be documented in some of our genomes.

We are the last of many types of humans—having displaced and absorbed all other hominin species that used to walk this Earth. If distinct groups of *Homo floresiensis* or Neanderthals were still with us today, we could not claim to be that different from all the other creatures. And the mystery of our special status, which baffled not only Wallace, would have been much more explicable through evolutionary processes. With the rest of the hominins all gone, we now appear to be vastly different from all other species. The current gap between humans and other animals—unbridgeable as it may appear— can be explained by standard, gradual change by descent with modification, along with the disappearance of intermediates.

If we continue to act as we have done, the prognosis is that we will also eradicate those species that are at present most closely related to us—the other apes. All great apes are endangered or critically endangered according to the International Union for Conservation of Nature. They are in that position because humans hunt them and destroy their habitats. And so these apes may soon join all the other long-gone species of *Homo, Australopithecus,* and *Paranthropus,* as beings of an ancient time, making our species ever more unusual on this planet.[88]

Though the extinction of the other apes remains the most likely scenario, we can, of course, conceive of alternative futures. We can use our compassion and our tamer tendencies to protect them. We can broadcast to others that our closest remaining relatives are headed to extinction and coordinate actions aimed at averting this outcome. If we accept moral culpability, we can plot a way forward that ensures unwelcome possibilities do not become reality.

———

Contrary to Wallace's conclusion, there are compelling explanations for how our minds, including the faculties that help us "transcend time and space," gradually emerged, and for how humans themselves created new selection forces that in turn accelerated the evolution of foresight and culture.

It has been a long journey from when our ancestors first had to eke out a living on the savannah. Away from the safety of the trees, there was selection pressure to cooperate, to innovate, and to prepare. Our forebears eventually relied increasingly on stone tools and started making them according to a plan even before they were needed. After hundreds of thousands of years, they innovated smaller stone tools, began to combine pieces together, and huddled around the fires they had learned to master. New evolutionary pressures emerged as humans began coordinating their goals, and, in time, communicated symbolically. The signs of modern human behavior began to leave lasting marks in the archaeological record: jewelry, burials, mobile containers, sophisticated weapons, cave art. The feedback loop between culture and foresight had finally gathered momentum. But we were not the only species exploiting this new niche. Competition with other upright-walking hominins may have introduced further selective pressures on foresight and cooperation. Certainly, the sheer fact of the demise of our cousin species is, in the end, the reason why today we appear so dramatically distinct from the rest of the animal kingdom.

Foresight was a prime mover in human evolution. The precise forces bringing about each of the critical steps in this journey are not easily established, but scientists are unearthing more and more evidence about how exactly our ancestors discovered the fourth dimension. Next we will turn to some of the tools humans invented to improve their foresight and bring the future ever more under their control.

CHAPTER 7

TRAVEL TOOLS

The true engine of reason, we shall see, is
bounded neither by skin nor skull.

—Andy Clark (1996)

If you return to the same spot every morning to watch the sun rise and every evening to watch it set, you will notice gradual changes. By placing markers where you see the sun cross the horizon, you can create a record of these changes over time and discover the annual pattern. About seven thousand years ago, some dedicated people near what is now the village of Goseck, in Germany, appear to have done just that.[1] They placed logs in two arcs, forty yards from the center of observation, which appear to have tracked the sun's rising and setting, until the sun's trajectory reversed and started to retrace the places it had earlier risen and set. These turning points are the winter and summer solstices, or the shortest and longest days of the year. With the annual pattern marked out, people could predict where the sun would rise and set tomorrow, and the next day, and in a month's time, and in a year's time.[2]

The Goseck circle is but one of many similar assemblies built in Europe beginning around this time, and among many examples of apparent

Figure 7.1. The reconstructed Goseck circle in the snow, with gaps marking the solstices.

astronomical structures found across the world. In Australia, for instance, the ancient Wurdi Youang stone arrangement appears to have been designed by Aboriginal people to align with the setting of the sun at the solar solstices and equinoxes. Similarly, a sixty-five-hundred-year-old stone circle in the Sahara, at Nabta Playa, seems to have marked the summer solstice and so the arrival of the monsoon season. Perhaps the most famous of such arrangements is Stonehenge in Wiltshire, England, which is around five thousand years old.[3]

After the end of the last ice age, humans were leaving signs suggesting they had followed the firmament and uncovered its long-term regularities. It wasn't only the sun they tracked either. An arrangement of twelve pits buried under Warren Field, a paddock near Aberdeen in Scotland, has been dated at about ten thousand years old and is argued to be the remnants of the world's oldest lunar calendar.[4]

People tracked stars and their constellations too. Some fifteen miles from Goseck, two treasure hunters in 1999 discovered a stash of long-forgotten artifacts with their metal detectors: a chisel, two swords, two axes, two spiral armbands, and a peculiar pizza-sized disk that had been carefully placed

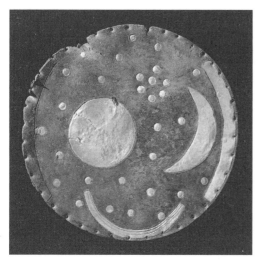

Figure 7.2. The Nebra sky disk. Two golden arcs attached to the sides, one of them now missing, appear to have represented the rising and the setting of the sun over the course of the year. Potentially the disk was a portable instrument for tracking winter and summer solstices.

upright behind the other objects.* Dated at over thirty-five hundred years old, the Nebra sky disk features a crescent, a large orb, and thirty-two smaller circles—all made out of gold—that seem to depict a waxing moon, a sun or full moon, and stars. A cluster of seven circles resembles the constellation Pleiades, which you may recall also appears in the Wurundjeri Dreamtime story of the fire-stealing Crow. It has been argued that the Nebra sky disk is the oldest known depiction of the cosmos.[5]

There were probably many ritualistic and other uses for such structures and objects. But one likely reason for this apparent obsession with the heavens is that tracking and predicting celestial patterns provides a tool for accurately foreseeing other, more down-to-earth matters.

Ultimately, we schedule much of our lives around the locomotive quirks of our planet as it meanders through space. Awoken by your alarm clock? It went off as the Earth crossed a specific rotational plane. Already looking forward to the end of the workweek on a Monday afternoon? You'll have to

---

* The two men recognized they had found something valuable and sold the whole find to a dealer the next day. This set in motion a chain of events worthy of a Hollywood movie, including multiple transactions on the black market for increasingly exorbitant sums. When the disk and other artifacts were offered up for the equivalent of hundreds of thousands of dollars in Switzerland three years after the find, the authorities finally swooped in, and the confiscated treasure was returned to Germany for scientific analysis.

wait for the Earth to rotate four more times before you can enjoy the weekend. Feel summer coming on? Your hemisphere of the Earth is becoming gradually more tilted towards the sun. Have an anniversary coming up? The Earth is in almost the same position relative to our star as the last time you celebrated. Whimsical as it may seem, your plans genuinely hinge on the movements of the space rock you live on.

It took humans many millennia of careful observation to unlock the secrets of celestial bodies and their relationship to Earth. After all, it wasn't until Nicolaus Copernicus published *On the Revolutions of the Heavenly Spheres*, shortly before his death in 1543, that people began to more widely appreciate that the Earth revolves around the sun rather than vice versa.[6] But even if it was commonly assumed that the Earth sits at the center of the universe, humans could still notice—and exploit—reliable patterns in the apparent motion of the heavens. By recognizing that the sun follows an annual cycle of rising and setting in the same places each year, they could accurately anticipate upcoming seasonal changes and coordinate their affairs.

Humans are of course not the only animals that adjust their behavior in line with the heavenly spheres. Some animals are so attuned to the sun that they completely lose their bearings in the rare event of a solar eclipse. When the shadow of the moon swept across Mexico on July 11, 1991, nocturnal bats left their roosts as the afternoon sky went dark at 1:23 p.m.[7] Orb-weaving spiders began dismantling their webs around the same time, only to begin hastily reassembling them as the sunlight returned minutes later.[8] It is important for animals to do the right thing at the right time, and so they have also internalized such regularities. There are many biological clocks ticking in nature: circadian rhythms regulate biological activity according to the twenty-four-hour solar cycle, and circannual rhythms regulate behaviors over the solar year, driving birds to migrate and chipmunks to hibernate.*[9]

---

* In one of the most remarkable cases, cicadas of the *Magicicada* genus have evolved to swarm in prime-numbered cycles, emerging in the springtime only once every thirteen or seventeen years. One theory is that these cicadas do this because their predators tend to have reliable boom-and-bust cycles where, say, every two or three years, there is a population explosion. If *Magicicada* had evolved a twelve-year cycle, therefore, they could have found themselves perpetually feasted on by predators hitting the peaks of their two-, three-, four-,

Many seasonal activities in humans, by contrast, are sustained by culture. We have not, for instance, evolved any instinctual means to know when to plow and when to sow. People must have known about regular seasonal changes for a long time, but figuring out exactly when these changes were most likely to occur would have been difficult without help. Calendars take out some of the guesswork. Ultimately, the increased predictive accuracy unlocked by early calendars drove further control over the future, as people could now imagine ways of coordinating ever more complex and large-scale cooperative endeavors to take place at particular moments in time. *Let's meet up again at the equinox* is a much better plan than *let's meet up again soon*.

Some philosophers even go so far as to suggest that our tools should be considered part of our minds.*[10] We are surrounded by—and often deeply dependent upon—clever innovations that bolster, prop up, and enhance our cognitive capacities. As we will see, much of what makes foresight so powerful resides outside our brains—and has done so for a very long time.

The world changes with the seasons, and it can be hard to eke out a living. By tracking regular patterns in landscapes, tides, animal migrations, and plant flowerings, people can predict and prepare for upcoming changes in food sources and hunting grounds. The Yolngu in Australia, for instance, divide the year into six seasons according to biological markers such as when the stringybark tree is in flower. These markers are directly tied to relevant activities, such as when to burn the grass or commence spear fishing. The Inuit in North America have also maintained oral calendars, organizing their affairs based on annual biological events such as when seals are born or caribou shed their antlers. Such traditions are possibly much older than the first signs of physical calendars.[11]

---

or six-year cycles. But by emerging and swarming every thirteen or seventeen years, they are less likely to face predator peaks (because no smaller numbers go into thirteen or seventeen).
* In a famous 1998 article titled "The Extended Mind," philosophers Andy Clark and David Chalmers introduced readers to a hypothetical man with Alzheimer's disease, Otto, who routinely uses a notebook to remember what to do where and when. Clark and Chalmers propose that Otto's notebook is part of his cognitive system, just as a neurotypical person's hippocampus is a part of theirs.

By passing on knowledge about the world, acquired through systematic observations of what happens where and when, people could populate their mental theater with more accurate sets, actors, and narratives. This meant they could foresee the opportunities and dangers that might lie ahead, in both space and time.

Aboriginal Australians, for instance, have long enhanced their capacity to navigate through harsh lands by singing songs that trace the path from landmark to landmark and resource to resource. By inheriting this knowledge, a person can envisage what to expect, even on journeys they have never taken before. Anthropologist John Bradley, who spent three decades living with the Yanyuwa people in northern Australia, reflected on his experiences with these *songlines* in his 2010 memoir *Singing Saltwater Country*. One of the Yanyuwa men with whom Bradley often spoke was Ron Rickett, who dictated a songline (kujika) in spoken-word form. The entire dictation is far too long to print here, but it describes a remarkable course through miles of difficult terrain—passing saltwater crocodiles, wallaby habitats, and plant life—on the way from Manankurra in Yanyuwa country to the scarce fresh water of the Walala lake. An excerpt reads:

> We are still following the path of the kujika and we turn eastwards and descend down to Wurrkulalarra creek and then we are in the water. We enter the waters of the creek and we sing the Blind Rainbow Serpent. We come to the sea, and we sing the sea grass, the tall tape grass, that sea grass is Rrumburriya, and the spotted eagle ray and then we turn and head north-north-west and sing the cabbage palms, the little corella and whistling kite. We keep moving northwards: we sing the spring waters and the file snake and the brolgas, and keep travelling. We near Walala; it is a big body of fresh water. We come to Walala on the west side; yes, at last we are meeting that place we call Walala.[12]

After listening to Rickett's narration and traveling the kujika himself, Bradley writes: "I began to see how kujika fixed the cartography of country in the consciousness of the individual." By memorizing the songline, the Yanyuwa people could learn the way from Manankurra to Walala, as well as the characteristics of significant animals and plants associated with each landmark along the way. Songlines thereby helped the first Australians acquire an encyclopedia of environmental knowledge on which they could draw to foresee the future better and so secure provisions in their vast and often unforgiving landscapes.[13]

With increased capacities to foresee the future in detail and awareness of periodic cycles in nature, people set out to manage and control their worlds. In some cases, by weeding out what they did not want to grow and nourishing what they did, people controlled the food that would be available to them in the future. Planting seeds only makes sense when you understand that they will turn into something else in the future, and after the last ice age, evidence for such behavior appears across the globe. From irrigating maize in Mesoamerica to aerating soil to grow bananas in the highlands of New Guinea, humans began to farm crops.[14] Likewise, people began to pen and care for livestock because they recognized the future payoff in milk, meat, or leather.

The rise of agriculture in the Fertile Crescent of the Middle East has been the most influential on the modern world, at least in the sense that descendants of many plants and animals domesticated there can still be found in our supermarkets today. The traditional narrative of cultural progress holds that, following the ice age, the inhabitants of this region readily abandoned a hunter-gatherer lifestyle once they discovered the benefits of agriculture. They settled down, tended their crop, and waited for their harvest. But this is too simplistic. The process would have been gradual, and not everyone who realized that agriculture could work would have wanted to become a farmer. Often there may have been little incentive for hunter-gatherers to willfully abandon their lifestyles, especially given the brutal, monotonous, and disease-ridden lot of the farmhand. In fact, as the political scientist and anthropologist James C. Scott argues in his book *Against the Grain*, some of the earliest city-states seem to have relied heavily on slave labor.[15]

Nonetheless, agriculture took over, eventually replacing hunter-gatherer lifestyles in most places that were suitable for farming. Data from the UN Food and Agriculture Organization show that around 50 percent of habitable land on Earth is now exploited for this purpose.[16] Our livestock account for around 60 percent of the biomass of all mammals on the planet, compared to 36 percent for our own bodies and a paltry 4 percent for all wild mammals combined.[17] Modern farming calendars have been cultivated to a fine art, with ideal planting times and field rotations for every crop and carefully scheduled uses of livestock for replenishing vital soil nutrients.

Stone circles and other apparent calendars appeared in Europe just after the spread of farming from the Middle East, with the Goseck circle constructed around five hundred years after agriculture reached the area.[18] But only with the emergence of writing do we have a clear record of calendars and their relation to agriculture. In Babylon, one of the early thriving metropolises of the Middle East, people began writing about their calendar some four thousand years ago. It comprised twelve months, each approximately the length of a lunar cycle, and also included a periodic leap month to account for the bothersome fact that twelve lunar cycles only equate to around 354 days. Babylonian astronomers tracked and recorded celestial events with precision and dedication. They even figured out that the relative positions of the Earth, sun, and moon repeat approximately every 223 lunar months, and they used this knowledge to accurately predict solar and lunar eclipses (similar to how the Antikythera mechanism did it).[19] The astronomers were not only observing the activity in the sky for its own sake, they were also discerning patterns in that activity and casting those patterns into the future for practical purposes here on Earth.

One of the most widely copied texts of Babylonian literature is known by the moniker MUL.APIN (named after the first two symbols inscribed on the cuneiform tablets). Archaeologists have uncovered more than sixty near-identical copies of this text, suggesting that it played a central role in Babylonian culture.[20] It contains a remarkable catalog of cyclical astronomical events, including the annual rising dates of dozens of stars and constellations—such as, of course, the Pleiades. This knowledge would have been invaluable to any society that had to carefully plan seasonal agricultural activities, as the Babylonians and their neighbors did. Indeed, several passages of MUL.APIN explicitly connect the sun and stars with farming activities, such as this imperative: "From the 1st of Simanu until the 30th of Abu the Sun stands in the path of the Enlil stars; harvest and heat."[21]

Figure 7.3. This clay tablet is one half of MUL.APIN. It is only 3.25 inches high and 2.4 inches wide. A person could have held it in one hand and read from it like a small smartphone.

On the other side of the planet, Mesoamerican peoples independently innovated another writing system, culminating in the Mayan script used on and around the Yucatan Peninsula from about two thousand years ago.[22] Tragically, the Spanish conquistadores burned nearly all Mayan texts before they could be deciphered. From an original catalog estimated in the thousands, the entire library is now survived by only four fragmented books made of the bark from fig trees, and some scattered phrases carved into stone.[23]

The most complete of the remaining Mayan books is the seventy-eight-page *Dresden Codex* (named after the German city where it eventually ended up). Much like the Babylonian MUL.APIN, this text is primarily dedicated to tracking recurring celestial cycles, including those of the sun, Venus, and Mars, and it also alludes to the Mayans' 365-day *Haab'* calendar.[24] As one conquistador recalled: "These priests . . . had books of figuras

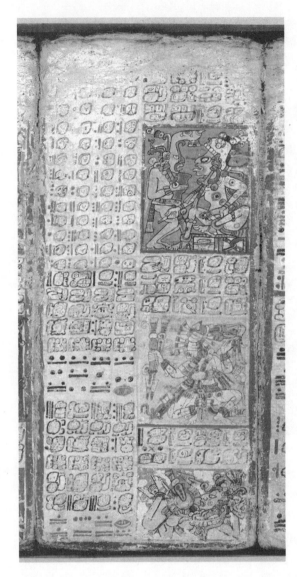

Figure 7.4. A page of the *Dresden Codex*.

by which they governed themselves and there they had marked the times when they had to sow and harvest."[25] Just like the Babylonians, the Mayans used their calendars to time their farming activities.

The benefits of establishing shared systems of time reckoning have been dramatic. As journalist David Ewing Duncan put it in his book *The Calendar*, humans around the world have long yearned "to calculate the movement of the sun, moon and stars, and to capture them all in a grid of small squares that spread out like a net cast over time: thousands of little squares for each lifetime."[26] Ever so surely, those little squares of time have become smaller, more refined.

———

Calendars based on whole numbers eventually go out of sync with the natural world. In 46 BC, Julius Caesar returned from war to discover that Rome, which had been relying on a 355-day calendar, had lost track of time. The harvest festival was being celebrated before the harvest.[27] Frustrated, he convened the best astronomers in the republic to fix the muddle. Thus the 365-day Julian calendar was born, with a leap day added every fourth year, and 46 BC became known as the *annus confusionis ultimus* ("the last year of confusion").[28]

If the Earth made precisely 365 turns every time it circled the sun, then calendars would have been a lot easier to construct and maintain. Instead, it makes an inelegant 365.2422 turns per cycle (and even that only approximately), which is why the Julian calendar included the leap year to get back on track. But, given that the number of annual Earth rotations isn't precisely 365.25 either, Pope Gregory XIII proclaimed another reform in the year 1582. The Gregorian calendar includes the rule that every hundredth year is not a leap year (no leap year in 1700, 1800, or 1900), except if that year is also a multiple of four hundred (so 1600 and 2000 were leap years).* If that's not puzzling enough, consider that the Earth's rate of rotation is also gradually shifting, meaning we sometimes have to add a leap second at the end of the year to account for minuscule changes in the length of a day. December 31, 2016, therefore, lasted 86,401 seconds, instead of the usual 86,400. *Happy New . . . Year!*

While calendars track events over multiple days, people have also innovated strategies to track events between sunrise and sunset. The Yaraldi in South Australia, for instance, used the movement of the sun to divide the

---

* As mentioned earlier, the moon takes about twenty-nine and a half days to circle the Earth (29.53 days to be more precise). This means the solar year is around eleven days longer than twelve lunar cycles, posing considerable problems for the early astronomers trying to align the two. No easy math here.

day into seven sectors: before dawn, dawn with rising sun, morning, sun at center, afternoon, sun going down, darkness.[29] These kinds of divisions may have emerged a long time ago. The movement of the sun, of course, is also conveyed in the changing shadows it casts, which people eventually began to track systematically. Sundials were commonplace around the ancient world, and the Egyptians may have even used the shadows of giant obelisks to visualize the passing of time.[30] Naturally, shadows have a considerable downside in that they do not show themselves in bad weather or at night. Also, unless you are right on the equator, the length of daylight constantly changes throughout the year and so do the shadows.

Clocks that measure time by dropping a consistent amount of some substance through a hole, such as sand or water timers, are far more reliable. These appeared in various cultures over the last four thousand years. MUL.APIN, for instance, includes several passages describing how the Babylonians carved up the working day and night using a four-hour water-clock cycle known as a *mina*.[31] Around the summer solstice, MUL.APIN specifies that "4 minas [sixteen hours] is a daytime watch, 2 minas [eight hours] is a nighttime watch"; whereas around the winter solstice, "2 minas [eight hours] is a daytime watch, 4 minas [sixteen hours] is a nighttime watch."[32] Early water clocks arguably reached their pinnacle with the prolific inventor al-Jazari, who built several marvelous time-keeping devices in the twelfth century, including an enormous clock in the shape of an elephant.[33]

Gear-based mechanical clocks were invented and gradually refined over the last thousand years, though it remains possible of course, as the discovery of the gear-based Antikythera mechanism demonstrates, that some civilization hit upon them much earlier and people merely lost the technology for some time.[34] The Dutch inventor Christiaan Huygens, who introduced the use of mathematical formulae to describe laws of physics, generally receives credit for building some of the first reliable clocks in the seventeenth century. Huygens's clocks were driven by the oscillating regularity of a pendulum and only out by a few seconds each day.[35]

The need for ever more precise measurement of time over the subsequent centuries—for instance, to accurately measure longitude while at sea—drove yet further innovations. By the twentieth century, the oscillation of quartz crystals was harnessed in digital clocks around the world, and the reliable resonance frequency of cesium eventually allowed scientists to build atomic clocks with pinpoint accuracy. Atomic clocks enable us to measure the minuscule difference between the time it takes light to travel from a satellite to Location A at one point and Location B only a few yards away—ultimately supporting GPS navigation. If the clocks in these GPS systems are out of sync by even a millionth of a second, this can translate into a discrepancy of over seven hundred yards on the ground.[36] In 2019, NASA launched their Deep Space Atomic Clock that will be off by only one second every ten million years.[37]

While the periods of twenty-four hours, sixty minutes, and sixty seconds may appear to represent arbitrary divisions of time within a day, there are good reasons we still use these numbers.[38] Both twenty-four and sixty are *highly composite*, meaning they can be divided into whole numbers in more ways than any smaller number. A twenty-four-hour day can be split into 2×12 hours, 3×8 hours, 4×6 hours, and vice versa; and a sixty-minute hour can be divided into 2×30 minutes, 3×20 minutes, 4×15 minutes, 5×12 minutes, 6×10 minutes, and vice versa. This makes for very convenient parceling of all aspects of life. "Eight hours labor, eight hours recreation, eight hours rest," or so went the nineteenth-century socialist slogan.[39]

The period of a week, of course, does not result from the division of a day but from the multiplication of days. People throughout history have introduced the concept of a week, typically to give workers at least one full rest day, and different cultures have used different numbers of days to define what a week means. The Edo in Benin used four days and the Javanese in Indonesia five days. The Akan in West Africa used six days, the ancient Romans eight days, and the ancient Chinese and Egyptians used ten days (as did the French from 1793 to 1802, at the height of the Revolution). The seven-day revolving week appears to have had its origins in Judaism around the sixth century BC.[40]

No matter how arbitrarily they may be defined, agreed measurements of time allowed humans to track events and schedule activities over both long and short timescales. It's all well and good to agree upon *what* you want to do and *where* you want to do it, but if you can't agree on *when* you want to do it, then many plans will quickly fall apart. So we bolster our flawed fore-casting abilities with tools that have made *when* an increasingly precise and verifiable matter.

————

The omens will be as favorable as I wish them to be.
—Julius Caesar (ca. 44 BC)

Tracking natural cycles and measuring events with precision clearly enables great powers of prediction and coordination. However, if you want more than that—if you want control over the future—it helps to develop theories about why some events follow others. These range from simple stories like *the ground is moist because it rained this morning* to complex narratives that aim to explain the state of the world with its jumble of fortunes and injus-tices. Causal explanations can improve predictions (*It has not rained, so the seedlings in the ground will dry out*) and open up opportunities to change the future in desirable ways (*If you want the plants to grow, keep watering them*).

Yet the true causal reasons linking events often fail to reveal themselves in any simple fashion. What brings about the rain? What causes plants to grow? How can we control rainfall and generate growth? Once an important co-occurrence has been noticed, whether the events were actually causally linked (*In a moist condition, many plants grew fast*) or not at all (*When we saw that bear, it started to rain*), people can attempt to influence future events (*Plant seeds where it is moist; worship that bear*).

The same was likely the case for many other attempts at controlling or influencing upcoming events. People may not at first have been able to dis-tinguish between the ritual of striking certain rocks in certain ways followed by fire and the ritual of dancing around a fire in certain ways followed by rain. Both may have been accompanied by elaborate rites and stories of gods and spirits. While it is plain for us now that one set of events is causally connected and the other is mere coincidence, these distinctions are not nec-essarily self-evident. It may not at all be clear, at first, that you need specific kinds of rocks to make a fire, and that they need to be in a specific state (wet ones won't work). Perhaps your rain dance did not work this time because you did not perform the ritual right, failed to sing the proper song, or did not make the appropriate sacrifice.

We often cannot foresee which ideas are gold mines and which are dead ends. Even one of humanity's most celebrated scientific minds, Isaac Newton, spent decades not only developing pioneering causal theories of mechanics but also pursuing the alchemists' dream of turning ordinary stuff into gold. (Before you snicker, note that, in partial vindication of the alchemists, it is in fact now possible to synthesize gold in nuclear reactors—though it is more expensive than the market price of mined gold . . . and tends to be radioactive.)

In 44 BC, Roman orator and wordsmith Cicero penned a peculiar tome titled *On Divination*, in which he considered the nature of cause, effect, and prediction. Written as a dialogue with his brother Quintus, the book opens: "It is an old opinion, derived as far back as from the heroic times, and confirmed by the unanimous consent of the Roman people, and indeed of all nations, that there is a species of divination in existence among men, which the Greeks call *mantikí*, that is to say, a presentiment, and foreknowledge of future events."[41]

For many centuries, the Romans' favored method of *mantikí* was augury: reading the future from the behavior of birds. True enough, birds flying south can signal the arrival of winter, but the Romans read much more into avian activities. Food falling from the beak of a sacred chicken was a good omen, whereas the same chicken refusing food was a bad omen. A raven appearing on the right side of the augur was a favorable portent, but a raven on the left side less so (and for crows the opposite rule applied). Gallic king Deiotarus always turned back from a journey whenever he saw an eagle in flight, and he believed this once saved him from staying in a room that later collapsed. At the height of antiquity, taking the auspices was serious business.

Quintus was an earnest defender of divination, believing that the future could be gleaned from bird behavior, cattle entrails, planetary activity, and lightning strikes. But Cicero himself was a staunch skeptic. He dedicated the entire second half of *On Divination* to systematically dismantling his brother's beliefs, and to arguing that any philosopher who endorsed divination should be "ashamed to utter such nonsense." Cicero's main contention was that everything must "find its cause in nature," and that the only plausible way for humans to accurately predict the future is to identify natural connections between events.[42]

Although the views of Quintus and Cicero might seem like polar opposites, they both ultimately spring from the same underlying cognitive capacity. Whether causal theories turn out to be natural or supernatural, they reflect people's attempts to understand and foresee the future better.

Recall that the Babylonian astronomy text MUL.APIN was largely an incredible protoscientific compendium and practical farming calendar. Yet the final section of this text includes peculiar passages such as "If the star of Marduk [Jupiter] becomes visible at the beginning of the year: in this year the crop will prosper" and "If a star flares up from the middle of the sky and sets in the West: a heavy loss will occur in the land." The omens of MUL.APIN's concluding segment all refer to uncertain future events like the fertility of newly irrigated land, the fate of horses and livestock, the occurrences of plagues and political revolutions, and the life spans of new rulers.[43]

Similarly, although the bulk of the Mayans' *Dresden Codex* contains meticulous records of the cycles of Venus and Mars, a large chunk is devoted to a 260-day divination calendar known as the *Tzolk'in*.[44] The codex contains dozens of Tzolk'in calendars, in fact, each bursting with distinct prognostications corresponding to every calendar day. Bishop Diego de Landa, who presided over a mass burning of Mayan books in 1562, recalled how such calendars were used: "The wisest of the priests opened a book and looked at the prognostics for that year, and declared them to those present and preached to them a little, recommending remedies [against the evils forecast]."[45]

A third major independent invention of writing occurred in China.[46] The earliest Chinese characters can be found on *oracle bones* dated to over three thousand years ago. Although some of these bones are etched with calendars, the vast majority appear to have been used for supernatural prophecies. Ancient Chinese diviners would write questions on a bone, such as "Will a toothache cause the king trouble?" and then burn the bone to produce cracks, which would be interpreted as foretelling the answer.[47]

Why didn't people recognize that celestial omens did not in fact correspond to future events on Earth? And how did the prognostications of MUL.APIN, the Tzolk'in, and oracle bones become so ingrained in their respective cultures? Cicero's supernaturally inclined brother, Quintus, had this to say on the subject: "Unfavorable auguries—and the same may be said of auspices, omens, and all other signs—are not the causes of what follows: they merely foretell what will occur unless precautions are taken."[48]

Unless precautions are taken! Under this view, it is easy to see why divination techniques might endure. If the unfortunate prophesized event does not occur, then congratulations, you took the right preventative action. But if the event does happen, then shame on you for failing to aptly heed the warning. In either case, you may at least find yourself reflecting on a possible future event—even if that possibility hadn't occurred to you before. And Quintus's superstition is hardly confined to the annals of antiquity. After the assassination attempt on President Ronald Reagan in 1981, Reagan's wife Nancy insisted on consulting a clairvoyant while putting together the presidential schedule. Based on the clairvoyant's advice, Reagan's chief of staff maintained a color-coded calendar, with "bad" days marked in red and "good" days marked in green. Written alongside January 20, 1986, for instance, was "nothing outside the WH [White House]—possible attempt."[*49]

Even if it does not work as advertised, divination is both a product of reflection on foresight and an attempt to enhance foresight. Humans routinely recognize they cannot predict the future with much precision, and so they

---

* During the nineteenth and early twentieth centuries, even distinguished scholars at the heart of scientific discovery, such as evolutionary theorist Alfred Russel Wallace, attempted to divine the future through séances, spirit boards, and other psychic techniques. However, when skeptics cast their expert eye over claims of supernatural ability, they uncovered nothing but smoke and mirrors. The magician and escape artist Harry Houdini offered prize money to anyone who could demonstrate supernatural phenomena—to no avail. Like a good scientist, Houdini remained open to the possibility of being proven wrong, and even set out to do so himself from the afterlife. He hatched a plan with his wife, endeavoring to send her a coded message from the beyond, and she duly held annual séances after his death. But he never showed up.

outsource their predictions to external tools and techniques, some of which work and some of which do not. But even the most impotent prediction techniques can at least get people thinking about diverse possibilities and trying to bend the future to their will.

————

With the invention of writing, people had a new tool to keep track of events over time and to share explanations about why the world is the way it is, as well as predictions about where it is headed. Modern research has illustrated how the ability to keep records may have also bolstered trading.

In one study, researchers introduced pairs of anonymous people, let's call the people Josh and Anna, to a trading game. Josh was given some money, and whatever proportion of this money he passed on to Anna would be tripled in value by the experimenters. In turn, Anna could then share some of the windfall back to Josh. The game lasted for ten rounds and was played by either just two people or by groups of ten people all sending money back and forth to each other. With ten participants, it was quite difficult to keep track of which trading partners were returning a fair share and which were not. Critically, however, in one condition participants had the option to keep a record of who sent how much money in each round, as well as how much was received in return. When they could keep records, participants were much more likely to split the money evenly, which in turn led to the other players sending more money to them in the future. In the long run, then, the record keepers ended up with much fatter pockets than the other participants.

The researchers speculated that in early human societies, record-keeping would have similarly enabled greater trust and coordination between strangers in complex trading networks. As the experiment demonstrates, record-keeping can ensure that any uncooperative behavior does not go unnoticed, which in turn creates an incentive to keep cooperating. Once such a cultural tradition of cooperation gets up and running, people can plan even large-scale trades that would have previously been too risky.[50]

Record-keeping doesn't necessarily need to take the form of full writing to be useful in this manner. Recall from Chapter 2 that the Sumerians originally encased tokens in sealed clay to keep track of debts and taxes.[51] Likewise, Chinese writing is thought to have undergone considerable development prior to the earliest evidence of oracle bones.[52] The *Tao Te Ching*, a petite book of Chinese parables and aphorisms from around twenty-five hundred years ago, alludes to a distant past when people used "knotting of rope" in place of writing.[53] Another ancient Chinese text, the *I Ching*, teases at the function of such ropes: "In great antiquity knotted cords served [rulers] for the administration of affairs."[54] No remnants of these mysterious Chinese knotted ropes have been discovered, but we do know that ropes were once used for the "administration of affairs" to great effect in at least two other parts of the world.

In an obscure passage published in a travel diary kept from 1821 to 1829, British missionary Daniel Tyerman recounted how the peoples of Hawaii kept track of property ownership:

> The tax-gatherers, though they can neither read nor write, keep very exact accounts of all the articles, of all kinds, collected from the inhabitants throughout the island. This is done principally by one man, and the register is nothing more than a line of cordage from four to five hundred fathoms [eight hundred to one thousand yards] in length. Distinct portions of this are allotted to the various districts, which are known one from another by knots, loops, and tufts, of different shapes, sizes, and colors. Each taxpayer in the district has his part in this string, and the number of dollars, hogs, dogs, pieces of sandal, quantity of taro, etc., at which he is rated is well defined by means of marks, of the above kinds, most ingeniously diversified.[55]

There are hints that people used ropes in a similar way on the Pacific islands of Samoa and New Zealand, and the Maori even tell a story about their ancestors transmitting messages between islands using ropes carried by a bird.[56] Linguistic evidence suggests that the modern-day inhabitants of Polynesia can trace their origins to a migration from Taiwan, and modern DNA evidence has revealed rare genetic markers shared by Polynesians and the Indigenous Taiwanese.[57] It is therefore plausible that the ancient rope record-keeping mentioned in the *Tao Te Ching* and *I Ching* had its origins in either mainland China or Taiwan, with the tradition continuing for thousands of years in the Polynesian islands.[58]

The most famous example of rope record-keeping, however, comes from South America.[59] Here, the Inka and their neighbors used knotted *khipus* for recording tax and property information, as well as—you guessed it—for keeping calendars.[60] Alas, the use of khipus gradually declined following the arrival of the conquistadores. Hernando Pizarro wrote that Inka accountants stood outside their storehouses knotting and unknotting their khipus to keep track of their depleting treasures as the Spanish looted them from within.[61] A few hundred remaining khipus can be found in modern museums and private collections, but the art of reading them has been lost.

We do not know if the khipu knots ever corresponded to the sounds of verbal speech as modern writing does. It might nonetheless be fair to draw

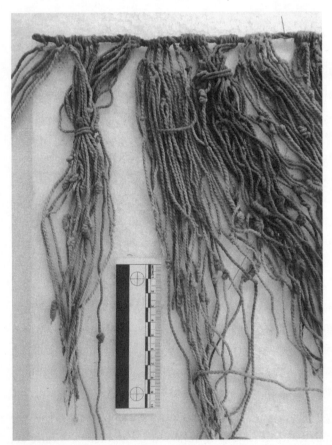

Figure 7.5. An Inka khipu recently excavated from the Inkawasi archaeological site in Peru.

parallels between khipus and the early cuneiform of Sumer, which originally included depictions of items and concepts rather than phonetic symbols corresponding to particular spoken sounds. Recall that the Sumerians gradually innovated cuneiform for similar purposes as the khipus served: keeping track of who owed what, and when the bill was due.

Yet gradually cuneiform began to incorporate more and more phonetic symbols, including some that seem to refer to specific individuals. The earliest readable name of a human being appears to belong to a bureaucrat known as Kushim. Sumerian accountants wrote Kushim's name on at least eighteen tablets from around five thousand years ago, including one inscription that has been interpreted as: "29,086 measures of barley to be delivered over 37 months to the government official Kushim, responsible for the brewery at the Inanna Temple in Uruk."[62] Cheers to that.

As historian Yuval Noah Harari writes in his bestselling book *Sapiens*, modern readers might find such texts "a big disappointment" because the early cuneiform script was not well-developed enough to express the philosophy, mythology, or laws of Kushim's time.[63] Still, just imagine how difficult it would have been to enforce a contract like Kushim's without some record that all parties could agree upon and refer to in the future. In the absence of such a document, thirty-seven lunar months (around three solar years) is perhaps long enough to conveniently "recall" that it was only *19,086* measures of barley owed. Early accounting records—clay tablets and khipus alike—bound parties to the terms of a trade across long spans of time, ensuring that large-scale, high-stakes cooperation between people with diverging interests could be enforced.

———

Eventually, the writing systems of the Fertile Crescent became sophisticated enough to express full phonetic sentences, allowing people to enshrine laws, treaties, and procedures that could enhance cooperation and order. This included classic texts like MUL.APIN and the more widely known Code of Hammurabi. Written around 1754 BC, the Babylonian code outlined the consequences people would face if they ran afoul of the law. Nearly every law in this code starts with the word *If,* followed by a potential future act and a state-mandated price to be paid.[64] "If a man put out the eye of another man, his eye shall be put out." So says Law 196.

Written rules promised that aggrieved people could test their cases against fixed standards of evidence, without having to resort to vigilante justice and risking escalating blood feuds. These rules made the consequences of anti-social actions patently predictable for everyone, and quelled unruliness by ensuring that scores would be settled by the state. Modern laws make even finer distinctions in their contingency plans for criminal behavior, often referring to the degree of foresight that preceded the crime. Recall the distinctions between intentional, reckless, and negligent wrongdoings.

Although parts of the Code of Hammurabi appear ruthless to modern sensibilities, others are perhaps more forgiving than you might expect. Consider the rules that protected debtors in case an unforeseeable "act of God" should occur. Law 48 reads: "If anyone owe a debt for a loan, and a storm prostrates the grain, or the harvest fail, or the grain does not grow for lack of water; in that year he need not give his creditor any grain, he washes his debt-tablet in water and pays no rent for the year." The code recognized that, even when both parties to a contract act in good faith when they agree on a future settlement date, sometimes things do not go to plan.

Writing assisted humans in resolving conflict and maintaining cooperation not only between individuals but also between groups. The oldest documented peace treaty is between the Egyptians and the Hittites from over thirty-two hundred years ago, ending two centuries of bloody conflict and committing each side to peaceful brotherhood.[65] It includes a farsighted obligation not to invade one another's land, even binding subsequent generations to the promise. It is known as the Eternal Treaty. Today, a copy is displayed at the United Nations headquarters.

Humans have long been aware that cooperation is fragile and have invented many other methods to make interactions more predictable and ensure the smooth functioning of their societies. They even innovated predictable ways to make outcomes unpredictable. Yes, they invented dice for gambling, but they also deployed randomization techniques to great practical effect. Aristotle's *The Athenian Constitution*, for instance, describes how such techniques can prevent corruption. Every male Athenian citizen above thirty years of age and in good standing was eligible to serve as a juror in civil and criminal proceedings. To have a chance of being selected for jury duty, however, he had to put a ticket into a lottery device known as the *kleroterion*. Jurors and magistrates were then assigned by lot to one of several courts. Aristotle wrote that the purpose of these randomization techniques was to make it "impossible for anyone to collect the jurors of his choice into any particular court."[66] The architects of this system were well aware of the potential for humans to pervert the course of justice and tried as best as they could to prevent this potential from manifesting in court proceedings.*[67] And because of the transparency of the process, an Athenian citizen could be confident that—should the time come—he would receive a trial in front of a random magistrate and jury of his peers.

Jurors did not serve on the Athenian court purely out of a sense of duty and good citizenship, however. They got paid. After hearing a case and placing an anonymous vote into an urn for counting, each juror received three *obols* made of copper or bronze.[68] Money itself was an important innovation that has enabled people to exchange goods and labor for the mere promise of a future payoff. As writer Jorge Luis Borges put it: "Any coin whatsoever . . . is, strictly speaking, a repertory of possible futures. . . . It can be an evening in the suburbs, or music by Brahms; it can be maps, or chess, or coffee; it can be the words of Epictetus teaching us to despise gold."[69]

Before money, trading required a *coincidence of wants*. If one person had an extra axe and another had an extra basket, and they each lacked what the other had, they could agree to swap. But often wants do not align that neatly. If one person wanted the axe, but the other did not want the basket in return, the first person might only be able to suggest bringing something else to exchange tomorrow. Money ensures such deals can be extended

---

* Anthropologist Pascal Boyer has argued that ubiquity of human divination can likewise be explained as a guard against vested interests interfering with collective decision-making. Many divination techniques involve some element of randomization—just think of those involving smoke, tea leaves, or tarot cards. Boyer suggests that such techniques may have emerged and survived because, if a decision is perceived to have been made by something out of peoples' control, then that decision is less likely to have been a result of corruption. Of course, people with vested interests can still manipulate what is being divined (especially when they are the only ones who can "interpret" the signs).

across time, and also enables exchange with third parties who weren't privy
to the earlier negotiation.

Though writing has widely facilitated money-based transactions, it is not
a necessary precondition. On the remote Pacific Ocean island of Yap, for
instance, people long used circular limestone disks—some weighing up to
four tons—to buy food or land and pay restitution.[70] Yap's economy was
based on a system where everyone agreed on who owned which stone at
any moment in time, and symbolic transfers of the largest stones were performed in public. In fact, some of these *Rai* stones sat in the same spots
on the island for centuries, all while their ownership changed hands many
times (figuratively speaking). Remarkably, the closest limestone deposit is
on the island of Palau, some 280 miles away over open water. For centuries,
people would make the long voyage from Yap to Palau, where they quarried
the stones, loaded up their boats, and headed back home. It was a perilous

Figure 7.6. Rai stones on the Pacific Island of Yap.

journey, and many a Rai stone found its ultimate resting place at the bottom of the Pacific. But the miners who were lucky enough to make it back with new stones had returned, as Borges would put it, with a repertory of possible futures in tow.

According to John Tharngan, the historical preservation officer of Yap, one of the most important factors in determining the value of a Rai stone was the number of lives lost on the journey to obtain it. Maybe so, but the stones only held value as money for as long as they could be exchanged for goods or services in the future.* Just consider that British twenty-pound notes still come printed with the words "I promise to pay the bearer on demand the sum of twenty pounds." Until 1931, a bearer could literally take the note to a bank and swap it for gold. Ultimately, the value of money—coins, notes, Rai stones, numbers on a screen—is based on the trust that people will continue to value it in the future.[71]

———

Is there a price for our extensive reliance on mental time travel tools like calendars, money, and writing? Socrates, the father of Western philosophy who famously never wrote down a word of his own, worried that if people learned to write, it would "implant forgetfulness in their souls. They will cease to exercise memory because they rely on that which is written, calling things to remembrance no longer from within themselves, but by means of external marks."[72] Nowadays, many people similarly worry about what will happen to human cognitive abilities as we become ever more reliant upon digital aids powered by the internet. "Is Google Making Us Stupid?" asked journalist Nicholas Carr in a widely reproduced 2008 article.[73] You might well wonder if your smartphone will shrink your hippocampus and downgrade your natural mental time machine.

---

* In the late 1800s, an Irish American sailor named David O'Keefe began exploiting modern technology to quarry his own Rai stones and exchange them with the Yapese. But O'Keefe's stones were valued less than the originals because they were obtained with far less danger. Nonetheless, as the influx of stones became greater and greater, prices on the island continued to inflate until the rock economy crumbled entirely. Today the Rai stones are typically exchanged for ceremonial purposes only, and the US dollar is circulated as Yap's primary currency.

These concerns may not be entirely unfounded. Some evidence suggests that extensively using GPS devices can lead to impairments in unaided spatial reasoning, and that storing information externally leads to an increase in false memories if the external store becomes unexpectedly unavailable.[74] However, by now it should be clear that Socrates, at least, was off the mark: literacy has not wreaked total havoc on cognition or society at large. And though it remains to be seen whether children growing up immersed in electronics, the internet, and artificial intelligence systems will be left hopelessly inept if the power cuts out, it is clear this digital frontier continues a long history of human reliance on offloading cognitive demands.

We have devised many means to enhance our capacities and compensate for our limits. We measure time, record history, and write contracts. These products of the human mind are all around us, creating a world—impenetrable to other animals—that governs what we do and how we do it. We borrow to pay back later, invest in the hope of returns, and buy insurance to hedge our bets. We save because we trust that we can cash in when the time is ripe. With writing we turn human language, already characterized by an open-ended capacity to refer to things removed from the here and now, from ephemeral sounds that immediately dissipate into fixed statements that can traverse the ages.

Our tools have dramatically altered our capacity to record the past, manage the present, and shape the future. Our minds are inextricably intertwined with our innovations, creating a labyrinth of promises and commitments underwriting the world we live in today.

CHAPTER 8

OUR SLICE OF TIME

We are at the very beginning of time for the human
race. . . . There are tens of thousands of years in the future.
Our responsibility is to do what we can, learn what
we can, improve the solutions, and pass them on.

—Richard Feynman (1955)

The human desire to predict and control the future has changed the world. From designing stone axes to bagging entire tool kits, from forecasting the seasons to cultivating crops, from planning trades to coordinating markets—foresight has been key to the human story. But it is by no means perfect, and we still get things terribly wrong.

To celebrate the start of 2020, a mother and her two daughters in Krefeld, Germany, wrote New Year's wishes on six lanterns and let them fly. The sight of slowly ascending sky lanterns has beguiled people through the ages. Yet these wishes for a brighter future reached the ape house of the Krefeld Zoo and set it alight—leaving dozens of primates, including two gorillas, five orangutans, and one chimpanzee, to suffer a horrible death in the ensuing blaze.[1] This tragedy might be read as a poignant metaphor for human powers—handed down by Prometheus the fire bringer—incidentally spelling disaster for other animals, including our closest relatives. Clearly, we have much to learn.

Human foresight will never be twenty-twenty. But we can compensate for our limits. As we have seen throughout this book, much of human power derives not from our outstanding foresight per se, but from our understanding of its strengths and weaknesses. We cannot foresee where our sky lanterns will land, and so it is for good reason they have been made illegal in so many countries.

This book has itself been an effort to grapple with the strengths and weaknesses of foresight. We explored the nature of our mental time machine, addressing what the ethologist Niko Tinbergen considered the four big questions about a trait: How does it develop? How does it work? How did it evolve over time? And what are its functions? In this final chapter we will recount the essential role foresight has played in recent history, showing that the disciplined application of foresight in the scientific method was a key to ushering in the Anthropocene. We will take stock of some of our successes and our enduring failures. Foresight is an awesome power. We will need to figure out how best to use it to navigate through this new era and secure a future worth looking forward to.

———

Recall that the scientific method essentially involves three steps.[2] Data must be gathered via observation or experimentation, potential explanations for these data must be generated, and, finally, hypotheses must be derived from these explanations and put to the test. Foresight is integral to this process: scientists are in the business of making and testing predictions. The genius of the scientific method is its built-in faculty for autocorrection. It forces scientists to look not for ways of proving their theories correct but for ways of ruling out alternative explanations. Scientists, like anybody else, may well get attached to their own ideas. But in the end, if the predictions are not consistently borne out, theories should be replaced or amended, leading to better theories that increasingly approximate the true workings of the world.

In principle, therefore, the more we uncover about these true workings of the world, the more we should be able to predict the future. In his foundational text of classical mechanics, *Philosophiæ Naturalis Principia Mathematica*, Isaac Newton laid out the first attempt at a complete theory of a universe where laws governing the motion of physical objects could be expressed with mathematical formulae—thus enabling perfect prediction.[3] If you know, say, the mass and acceleration of a moving object, you can multiply those two factors to calculate the force that object will exert in an upcoming collision. Together with his law of universal gravitation, Newton's three laws of motion captured the predictable yet complex trajectories of the planets, which the Babylonians and Mayans had begun to document millennia earlier. The publication of Newton's magnum opus on July 5, 1687, would prove to be a keystone of the Age of Enlightenment.[4]

The Enlightenment was a period of European history characterized by rapid accelerations in the dissemination of scientific findings and philosophical debate. It heralded a transition in the way that people sought to understand the world and themselves, built upon a foundation laid in various other times and places. Centuries before the scientific method was established in Europe, for instance, Arab scholar Ibn al-Haytham was already using observation, prediction, and testing to investigate the properties of light and vision.[*][5] Whenever it crops up, scientific thinking creates new avenues for foreseeing the future. Robert Hooke, Newton's contemporary and rival, was among the first to appreciate how this might be used to drastically improve human life. Hooke ventured that, one day, with tools built by the scientific method, "we may be able to see the mutations of the weather at some distance before they approach us, and thereby, being able to predict and forewarn, many dangers may be prevented, and the good of mankind very much promoted." Similar realizations about the predictability of nature, and the benefits this predictability might afford, occurred in various other domains.[6]

Take the history of predicting the tides. People living at the water's edge would have long observed its somewhat repetitive variations, but tidal

---

* Ibn al-Haytham also expressed the ideals of skepticism that would become a hallmark of scientific thinking: "The duty of the man who investigates the writings of scientists, if learning the truth is his goal, is to make himself an enemy of all that he reads. . . . He should also suspect himself as he performs his critical examination of it, so that he may avoid falling into either prejudice or leniency."

Figure 8.1. William Thomson's hand-cranked tide-predicting machine, 1876.

dynamics are not easily discerned. By a couple of thousand years ago, at least some people, including Seneca, had figured out that the swell of the tides could be roughly predicted based on the phases of the moon.[7] Yet it was only in the seventeenth century that Newton produced a reasonably reliable mathematical lunar theory of the tides.[8] This theory could have helped calculate tidal activity with near-perfect accuracy—if only the Earth had no continents or undersea topography. Later theories accounted for geography, enabling the construction of sophisticated devices like William Thomson's hand-cranked tide-predicting machine, which was built in the 1870s and incorporated the influence of ten different variables.

In October 1943, oceanographer A. T. Doodson received an urgent letter from the British Royal Navy containing eleven pairs of tidal variables for the mysterious "Position Z." The letter asked Doodson to provide hourly tidal predictions at this location for a four-month period commencing on April 1, 1944. He dutifully plugged the eleven variables into his tide-predicting machine and sent the results back to the navy. It turned out the Allies wanted these data to plan the D-Day invasion of Normandy, which, based in part on Doodson's calculations, was ultimately scheduled for the sixth of June.[9] The gradual refinement of tidal theories—from basic rules of thumb, through simple accounts incorporating a few variables, to complex models that have facilitated the detailed planning of large-scale operations—illustrates how humans have applied the scientific method to predict the future with ever-increasing precision. Today, wherever you are in the world, you can simply surf the web to find out when the next high tide will be.

We are now able to imagine and bring about future events that would have been unthinkable in centuries past. In 2012, an international consortium of physicists announced that experiments at the Large Hadron Collider at CERN, in Switzerland, had enabled the long-awaited discovery of the enigmatic Higgs boson.[10] No signs of this subatomic particle had ever been detected, and yet thousands of scientists and dozens of institutions were willing to pour millions of hours and billions of dollars into the search for its existence. Why? Because the math predicted it would be there.

————

Is everything predictable? In 1814, nearly forty years after publishing his own influential tidal equations, mathematician Pierre-Simon Laplace articulated an extreme version of Newton's lawful vision of nature. Laplace imagined an omniscient being who knew "all forces that set nature in motion, and all positions of all items of which nature is composed." Laplace concluded that, for such a "Demon," "nothing would be uncertain and the future just like the past would be present before its eyes."*[11]

Laplace's thought experiment implies that the only reason humans cannot predict future events with certainty is our ignorance about the relevant

---

* Although Laplace gets the credit for this idea, he was in fact beaten to the punch by nearly two thousand years. In *On Divination*, Cicero wrote: "If there were a man whose soul could discern the links that join each cause with every other cause, then surely he would never be mistaken in any prediction he might make. For he who knows the causes of future events necessarily knows what every future event will be." Perhaps Laplace's Demon should be known as Cicero's Demon.

factors involved—and not because anything in the future is fundamentally unpredictable. But Newton himself had been a little more circumspect. Although he concluded the preface of the *Principia* by asserting that all of nature might follow deterministic mechanical principles, he nonetheless acknowledged that he might be wrong, and that this hypothesis may eventually give way to "some truer method of philosophy."[12] Perhaps the future was not entirely predetermined after all.

That truer method of philosophy arguably arrived in the early twentieth century, as new theories of quantum mechanics began to place hard limits on even Laplace's Demon when it came to predicting the future. The Copenhagen interpretation of quantum mechanics, for instance, supposes that even with perfect knowledge of a particle's past position and momentum, it is impossible to perfectly predict both its future position and momentum—and that each of those variables remain fundamentally uncertain until they are measured. Albert Einstein disagreed with such interpretations, famously claiming that God "does not play dice."[13] Indeed, some interpretations of quantum mechanics, such as Louis de Broglie's pilot wave theory, maintain a deterministic view by presuming the existence of variables currently inaccessible to humans.[14] The debate continues, at least in part because competing interpretations of quantum theory lead to exactly the same predictions.

Just as the scientific method of observing, predicting, and testing can be applied to the motion of objects through space or the activity of subatomic particles, so too, of course, can it be applied to the chemical reaction of nitrogen and oxygen or to the life cycle of a frog. B. F. Skinner argued that understanding human behavior was no exception: "If we are to use the methods of science in the field of human affairs, we must assume that behavior is lawful and determined."[15] The regularities this approach has unearthed in psychology, however, are not quite as regular as those that have been distilled in chemistry or physics. One common-sense argument for this imprecision is that people, unlike particles, appear to have the capacity to make their own choices, driven by unique considerations, plans, and ambitions.

So it may turn out that not all is predetermined, and that real randomness or free-willed agents can truly make the future different from what could have been foreseen.[16] Certainly, when it comes to the whims of people, the unfolding of current affairs, fashions, sports, economics, or geopolitics, our capacity to predict leaves much to be desired. Still, the last few centuries have witnessed rapid increases in our scientific understanding across many disciplines, from astronomy to zoology, and this understanding has enabled ever greater prediction of ever more aspects of the future.

With greater scientific understanding, people also gained increasing control over the future, setting humans on a course of radical technological upheaval. The Industrial Revolution—with its new steam engines, coal mining techniques, and textile factories—was arguably made possible by a widespread appreciation of Newton's mechanistic philosophy by aristocrats and tradespeople alike. As sociologist Jack Goldstone put it: "In strongholds such as the Royal Society . . . scientists, engineers, and entrepreneurs [came] together to learn mechanics and discuss how this knowledge may be applied to improve production and society."[17] The notion of an intimate relationship between science, technology, and "progress" quickly spread—as did careless pollution and dreadful working conditions, not to mention slavery, colonial exploitation, and warfare with ever more sophisticated arms. The innovations rolled on, bringing electricity, internal combustion, telecommunication, and, eventually, microchips, satellites, and weapons of mass destruction.

For better or worse, science and technology have transformed the world. The population of *Homo sapiens* has exploded since the Industrial Revolution, from about one billion people two hundred years ago to some eight times that figure now. We far outnumber all the other primates combined—great apes, small apes, monkeys, lemurs, and the rest. Of course, our numbers are dwarfed by those of insects, let alone bacteria. (By current estimates, there are some forty trillion bacteria in your body alone—arguably making this really the planet of the germs.) We also fall short when it comes to biomass, with the world's minuscule viruses together weighing about three times that of all humans combined. Even worms outweigh us by some three times, fish well over ten times, and arthropods by fifteen times.[18] Still, we have an outsize impact. The mammals now most common on the planet are those we farm. As noted in the previous chapter, only 4 percent of mammalian biomass, including that of all the giants such as elephants and whales, can be found in the wild. The rest is made up of humans and livestock under our control. Only 30 percent of all the bird biomass flies free; 70 percent is farmed poultry. And our impact on Earth is not restricted to its organisms, with our split atoms, forged steel, and synthesized plastics. The combined weight of human material products (our buildings and roads, our computers and light bulbs, our trash) has been estimated at thirty trillion tons—or approximately 66,000,000,000,000,000 pounds.[19] Welcome to the Anthropocene.

———

Let's briefly look at a few of our apparent successes before turning to our continuing failures and challenges on the horizon.

Recent changes to our technological tool kit have been extraordinarily rapid. Take flight, for one example. There were only a few decades between when the Wright brothers designed the first powered aircraft and when mathematicians such as Katherine Johnson calculated the flight paths for spacecraft to orbit the planet and for Apollo 11 to land on the moon. In 2021, NASA completed the first remote-controlled flight on Mars, which likewise involved meticulous planning to account for the sixteen-minute signaling delay from Earth. These dramatic advances occurred in the span of one human lifetime: the world's oldest person at the time of writing—Kane Tanaka—lived through all three of these aeronautic milestones.

Thanks to advances in medicine, as well as in related domains such as hygiene, safety, and public health, babies born today can expect to live about twice as long on average as those born merely a century ago. And visiting the doctor has typically become far less terrifying. Imagine facing a mastectomy, as Fanny Burney did, without anything to dull the pain. Even kings and queens at any point bar the last hundred years or so would not have had the life expectancy ordinary citizens of most countries have today. They would have suffered operations without anesthetics and died spluttering with common disease like everyone else. As recently as the eighteenth century, five reigning European monarchs died from smallpox. Modern medical science has given humans new control over their own biology: the capability to heal injuries and cure diseases, and even to prevent problems before they arise.

Through strategic immunization, humans have overcome once devastating maladies such as smallpox and polio. The discovery of organisms not evident to the naked eye—bacteria, viruses, yeast—opened new opportunities for understanding and control.[20]

We have gained considerable control not only over survival but also over reproduction. With mental time travel, people must have long understood that sex leads to babies; and with modern science we have been able to increasingly break this link.[21] Contraceptive technologies like the pill and fertility treatments like IVF enable people to have sex without making babies and make babies without having sex.

A key driver of modern technological advances has been the ever more efficient exchange of information. Instead of dealing with each of life's challenges afresh, we use writing to amass and share a vast arsenal of knowledge. The invention of the printing press significantly accelerated the reliable dissemination of knowledge and so played a key role in the scientific revolution and the Enlightenment. Later, the discovery of electromagnetic waves opened up new worlds of communication over distances of space and time: telegraph, radio, television. Today, the internet gives people instant access to an astonishing wealth of data that has been accumulated by millions of other people, including scientific findings, maps, itineraries, do-it-yourself instructions, and reviews. We can find out what happened to others who took a certain route, what challenges they faced, and how these could be overcome. When we then confront our own problems, we need not fly blind but can draw on the wisdom of the crowd.

Much of our apparent progress has been made possible by people foreseeing a better world, communicating about it, and cooperating to create it.[22] Just consider recent developments in our coordinated efforts to anticipate and prepare for natural disasters. On December 26, 2004, an earthquake in the Indian Ocean triggered a tsunami that brought devastation to the shores of Southeast Asia, killing over two hundred thousand people and displacing millions more.[23] In the wake of the tsunami, UNESCO established a warning and mitigation system, measuring seismic activity to rapidly alert authorities.[24] Early-warning systems now also include buoys tracking changes in water levels, and researchers are working on other advanced indicators—for instance, using hydrophones to measure acoustic waves. Once the threat is registered, warnings aim to reach everyone who might be affected through a host of channels: personal emails and SMS messages but also sirens or loudspeaker announcements. Today, tsunami warning systems around the world are maintained to give people the opportunity to rapidly execute evacuation plans and mitigate the destructive impact of these disasters. Even if none of this is foolproof, we are now much better prepared than we once were.

Thinking ahead has played an essential role in human flourishing and has helped to bring us plenty we can be thankful for. But now is not the time for smug self-congratulations.

————

Foresight often fails. We still frequently trip up and miscalculate even the immediate turn of events (which has kept *America's Funniest Home Videos* in business for decades). Our slightly longer-term predictions, such as our plans for a relaxing holiday, continue to be upended by unforeseen hurdles: hailstorms and rude hotel managers, traffic jams and pilot strikes, acid reflux,

Figure 8.2. Approximate locations of deep-ocean tsunami detector buoys providing real-time warnings. Locations drawn based on Bernard & Titov, 2015.

twisted ankles, lost bags—not to mention global pandemics. Nor do we reliably predict the positive: new cuisines, new hobbies, new romances. It sometimes feels like our foresight is awfully blurry.

Even with access to vast amounts of information and with the help of extraordinarily sophisticated tools and technology, we are frequently well off the mark. Scientists, in spite of their training, may confuse correlation with causation, or draw too firm a conclusion when data are in line with what they hoped for. The folly of the crowd can mislead and confound: echo chambers that confirm our presumptions, to say nothing about targeted disinformation campaigns and deepfake videos, can bias our reasoning and predictions.

The public arena is packed with leading figures failing to foresee what seems obvious in hindsight. Five years before Horace Wells first used nitrous oxide to quell the pain of a dental operation, the renowned surgeon Alfred Velpeau claimed that "the abolishment of pain in surgery is a chimera" and pointless to pursue. At the end of the nineteenth century, William Thomson, whose tide-predicting machine we lauded earlier, predicted that airplanes would be impossible and that radio had no future. Einstein asserted in 1932 that "there is not the slightest indication that [nuclear] energy will ever be obtainable," whereas Alex Lewyt, the president of a vacuum cleaner company, foretold in 1955 that "nuclear-powered vacuum cleaners will probably be a reality in 10 years." The US postmaster general Arthur Summerfield pronounced in 1959 that "before man reaches the moon, mail will be delivered within hours from New York to California, to Britain, to India or Australia by guided missiles." When humans did first land on the moon, many people predicted there would be lunar colonies by the end of the century, with Venus and Mars ripe for further waves of colonization. Yet few people foresaw what was actually going to transform much of our lives: innovations such as the internet and smartphones.[25]

A failure to foresee can have treacherous consequences. To smooth the operation of car engines, the inventor Thomas Midgley Jr. introduced lead to gasoline, which he did not anticipate would turn out to produce one of the world's worst pollutants. Nor did he foresee that the CFC (chlorofluorocarbon) he introduced to refrigerators would be a major cause of ozone depletion. As environmental historian J. R. McNeill put it, Midgley "had more impact on the atmosphere than any other single organism in earth history."[26] Evidently, many of our innovative solutions to problems create new problems that require further solutions. Midgley's tragic story ended, true to his luck, when he was strangled by one of his own inventions (a pulley and rope system designed to help him out of bed).

In many cases the potential for disaster should have been easy to anticipate. The Krefeld Zoo sky lantern tragedy is unfortunately no anomaly. Take Balloonfest '86, when a charity organization in Cleveland, Ohio, attempted to set a Guinness World Record by releasing 1.5 million helium-filled balloons into the sky. After six months of careful planning, thousands of people gathered on a Saturday afternoon to prepare for the spectacle. Local children excitedly filled and tied balloons for hours in the sun until their fingers blistered. Then, at around two in the afternoon, the balloons were released in the public square to much fanfare. It seems unthinkable in hindsight that this stunt was allowed to go ahead—it's as if nobody involved could anticipate the now obvious consequences that were about to unfold. Soon, the city and surrounding areas were inundated by the descending garbage; thousands of balloons drifted onto the city airport's runway and grounded air traffic, while thousands more wreaked havoc on the roads. Two lost fishermen drowned when coast guard rescuers could not spot them among all the balloons bobbing around in Lake Erie.

How many balloons have we let loose to celebrate birthdays, weddings, and other events, let alone citywide publicity stunts? It is difficult to maintain that soaring helium balloons are merely an innocent delight for children once we apply our understanding of object permanence and admit that they do not disappear as they glide out of eyeline. Single-use plastics, from straws to forks, have transported delicious treats into our mouths for decades—but their widespread use has many unintended costs.[27] The UN estimates that one million plastic bottles are purchased every minute and five trillion plastic bags are used every year.[28] So one of our greatest farsighted innovations, the mobile container, now carries a dramatic environmental downside. Plastic waste is everywhere, from the depths of the Mariana Trench to the top of Mount Everest.[29] The Great Pacific Garbage Patch, an enormous diffuse collection of debris and microplastics, litters ocean waters from Japan to the USA.[30]

By and large, we can no longer claim ignorance about the environmental consequences of our actions. Continuing to litter, to emit, and to destroy is reckless, and often simply driven by a focus on immediate monetary gains and a willful neglect of factors outside of market forces. Humans tend to discount the future, after all, prioritizing short-term benefits at the expense of long-term costs. Those costs, of course, may end up being borne by other people. Even other species. Just consider the five hundred or so primate species on our branch of the tree of life alone. Some 60 percent of them are threatened by extinction as a result of human activity, not just through hunting but primarily as a side effect of destroying their habitats by logging, drilling, farming, building, dumping, damming, and mining.[31]

Remember that our close hominin relatives have already gone extinct, perhaps with some help from our own ancestors. If we continue on this path, all the apes will also be extinct in no time, leaving the remaining monkeys to be our closest living relatives. Our own descendants may then have even more reason to marvel at how different they are from the rest of the animal kingdom.[32]

Of course, the conditions for life on Earth have always been in flux. Continents shift, the climate changes, species go extinct. There have been five mass extinction events, with the last one following the asteroid collision that appears to have wiped out the dinosaurs some sixty-six million years ago. None of these events were caused by a single species running amok, but the root cause of the current calamity is undoubtedly us.[33] As the journalist Elizabeth Kolbert laments in her Pulitzer Prize–winning book *The Sixth Extinction*, "We are deciding, without quite meaning to, which evolutionary pathways will remain open and which will forever be closed. No other creature has ever managed this, and it will, unfortunately, be our most enduring legacy."[34]

The ship needs steering. As the only animal on the planet capable of foreseeing the long-term consequences of their actions, we have choices faced by no other creature. Our farsightedness burdens us, and us alone, with responsibility.

Let's look at our current ecological problems as an example of how we might put our foresight to better use. Until the beginning of the nineteenth century, people did not even realize that species could go extinct at all, let alone be driven to extinction by our own actions. In spite of observations such as the disappearance of the dodo on Mauritius within a century of its discovery, the idea of extinctions was not properly established as a fact of life until naturalist Georges Cuvier described fossils of what he eventually called the mastodon. By carefully comparing the fossil anatomy to that of living elephants, Cuvier showed that these fossils represented a distinct species—way too conspicuous to have been overlooked if they were still alive. Only by recognizing the possibility of extinctions can we plan to avoid them.[35]

Today we know that even the seemingly endless oceans are in dire need of careful planning to avoid wholesale extinctions caused by human activities.[36] In 2010, the United Nations Strategic Plan for Biodiversity set a target to protect at least 10 percent of the world's oceans.[37] But not all ocean waters are the same. Australia's waters, for instance, contain the largest reef system in the world, where some six hundred types of corals have created some three thousand reefs over 130,000 square miles.[38] The Great Barrier Reef is home to an extraordinary diversity of life, including whales, dugongs, and turtles, as well as hundreds of species of sharks, rays, and birds, not to mention thousands of species of other fish and mollusks. The region is also home to fishing operations, industrial ports, and a sizeable tourism industry, raising important questions about how to balance conflicting conservation and economic goals. The former chief scientist of The Nature Conservancy, Hugh Possingham, and his colleagues developed the Marxan algorithm—a systematic approach to spatial conservation planning—that has been used to rezone the Great Barrier Reef.[39] This approach recognizes diverse bioregions and the need to impose distinct *no-take* zones to protect a bit of each type. It was the first large-scale, systematically designed reserve system incorporating huge amounts of biological and economic data to boost conservation outcomes. Now over 120 countries have used this planning software and approach to build their land or sea protected-area systems.

Rather than being driven by intuitions or emotional appeals about dolphins, pandas, tigers, and other charismatic species, such approaches mean we can now systematically analyze the likely costs and benefits of alternative courses of action. Ultimately, much hinges on what we value and want to achieve, confronting us with tough moral decisions. Science might be able to help us look ahead, but it is up to us to choose which future to pursue.

After whaling decimated the number of humpback whales migrating between Antarctica and the Great Barrier Reef to a few hundred, the International Convention for the Regulation of Whaling helped their numbers to gradually recover, and now these seas are again teeming with these cetaceans.[40] We can learn from our mistakes. After all, balloons are no longer allowed to be released for celebratory purposes in Cleveland, and the European Union finally outlawed single-use plastics in 2021. Every country in the world has stopped using leaded fuel, a hundred years after Thomas Midgley introduced it. Following the discovery of the hole in the ozone layer, people across the world phased out the use of the manufactured chemicals responsible—including Midgley's refrigerant—and managed to initiate a recovery. The ban on chlorofluorocarbons has been ratified by all countries, and consumption of ozone-depleting substances has fallen to less than 1 percent of what it was in the 1980s.[41] In light of the rapid increase in global temperatures caused by greenhouse gas emissions, the 2016 Paris Agreement commits governments the world over to actions aimed at keeping global warming to below two degrees Celsius compared to preindustrial levels.

These concerted global efforts are extraordinary achievements in recognizing our mistakes, exchanging our forecasts, and plotting a way out of our troubles. Mind you, even with cooperation negotiated and commitments made, it can all come to naught if we do not actually follow through.

For the last ten thousand years or so, our planet has been in a relatively stable state known as the Holocene. But our recent activity has affected this equilibrium, bringing about the Anthropocene. We are facing some irreversible tipping points: climate change, atmospheric aerosol loading, ocean acidification, mass extinction.[42] Things we have taken for granted about nature—the animals and plants, the rain and the seasons—may radically change unless we behave more sustainably. Imagine what the world will be like if we do not drastically reduce activities we know to be harmful, such as extensive carbon emission, deforestation, and plastic pollution. If we do not implement sweeping changes to the energy sources we rely on, do not protect or restore habitats, and fail to build sustainable economic systems, we may yet yearn for the relative stability of the Holocene.

———————

> *It is a highly necessary part of foresight to be*
> *conscious that everything cannot be foreseen.*
>
> —Jean-Jacques Rousseau (1762)

Given how high the stakes are, we better figure out how to wield our foresight effectively and avoid its pitfalls. Recall that reflecting on foresight can give us an edge. So before we return to the big challenges humanity faces, let's take a closer look at some of the limits and shortcomings of our foresight and how we can try to compensate for them.

By some estimates, 80 percent of the population hold unrealistically optimistic beliefs about the future. This optimism bias means we tend to underestimate the chances of negative events and overestimate the chances of good ones.[43] Most people think they are less likely than they really are to be in a car accident, fall victim to a credit card scam, or get divorced and more likely to win an award, have gifted children, or live to a ripe old age. And these beliefs are not just a glass half full of lighthearted positivity. Overly optimistic forecasts can lead to gung-ho medical decision-making, overzealousness in military operations, and financial bubbles as investors get caught up in excitement about their gilded prospects.[*44] The cognitive neuroscientist Tali Sharot and her colleagues have documented that optimistic beliefs

---

* Many cons have been developed to take advantage of optimism. Probabilities can be used to create compelling schemes, such as the one where the con artist sends different predictions—say about stock market changes or the outcome at the races—to many people. Some of these forecasts, by chance, will be correct. The trick is to then follow this up with another round of predictions but send these only to those recipients who had received the winning prediction earlier. Do this a handful of times and those who happen to have received a 100 percent track record of correct predictions will think there is a system that really works—and may well be willing to pay handsomely for the next prediction.

are maintained in the face of reality because people tend to downplay bad news when forming their estimates of what the future holds (such as that it's much more likely that you will die of cancer than you assumed), all the while happily embracing good news (like that your chances of falling off a ladder are only half as high as they seemed at first blush). Even simply imagining a future event again and again can make it seem more likely, which can also contribute to unrealistic optimism when people fantasize about brighter possibilities on the horizon.[45]

In Nobel Prize–winning work, psychologists Daniel Kahneman and Amos Tversky documented many quirks of human prediction and decision-making worth bearing in mind. Consider the planning fallacy: people have a tendency to predict that plans will be executed more quickly than they typically are. Knowing about this fallacy, we can add a buffer to estimated time frames, bumping up initial forecasts whether for completing DIY jobs around the house or for the construction of a national infrastructure project.[*46] Or take the availability heuristic: people often simply estimate the likelihood of a possible event based on how easily it comes to mind. If easy to imagine, it must be likely; if difficult to imagine, it must be unlikely. This heuristic sometimes works well because events that occur frequently are encountered often and are therefore easily imagined. However, scenarios can also spring readily to mind for reasons other than their frequency—for instance, because something similar occurred recently. Events that are vivid, distinctive, or shocking—like a startup company hitting the big time or a plane crashing—tend to be effortlessly envisaged and so are predicted to be more likely to occur to us than they really are.

When predicting future events, people often neglect base rates: the typical frequency of the event occurring.† By contrast, political scientists

---

* The buffer itself, of course, may also be subject to the planning fallacy, leaving you to underestimate how much extra time you might need to actually get the job done. This in turn could lead to an infinite regress. Consider Hofstadter's Law: "It always takes longer than you expect, even when you take into account Hofstadter's Law."

† Base rates are important and easily overlooked. Let's say you tested positive to a disease. No test is perfect and you are told this one has a false positive rate of 5 percent. This does not mean that you have a 95 percent chance of having the disease, as one might assume. If the disease occurs in, say, 0.1 percent of people, then the actual probability that you are

Philip Tetlock and Dan Gardner found that "superforecasters"—people who excel at accurate prediction—tend to start with such base rates and then adjust their predictions after drawing on all the relevant additional information they can get their hands on.[47] So if they were to assess the probability of their plane crashing, they would begin with the low base rate of how often planes typically crash and then take into account other factors such as a flight path over a war zone or an ominous weather pattern. But "superforecasters" are not superhuman. We can all improve our forecasting by starting with base rates, adjusting likelihoods given special circumstances, and frequently asking ourselves whether we might have overlooked something or whether our assessment reflects any personal biases. Studies of such methods have primarily focused on predicting geopolitical events like next year's global unemployment rate or the outcome of a presidential election, but the insights apply across the board.

Consider how long you can expect to live. In trying to figure it out, one option is to simply take into account the average age at death of people in general. In 1693, Edmund Halley (of comet fame) made one of the earliest attempts to estimate life expectancy based on systematic demographic records.[48] Today, you can consult vast publicly available population data. According to the UN World Population Prospects, the average global life expectancy in 2019 was 72.6 years. Now, if you don't know anything else, the average is the best prediction. But, of course, you know a lot more relevant information: how long you have already lived, your sex, whether you smoke, where you live, and so on. This lets you fine-tune your estimate, adding a few years because of certain facts, deducting some for others. Even so, a new invention, outbreak, or discovery can quickly upend expectations about your demise. In spite of the most considered calculation, you may find

---

sick if you test positive is merely 2 percent. Chances are, you'd be fine. When we unpack the problem in terms of relevant frequencies, it becomes easier to see why. A base rate of 0.1 percent means one in one thousand people is expected to be sick, and so should test positive. But given the false positive rate (5 percent), you would also expect fifty out of one thousand people to test positive even though they are not sick. So of one thousand people tested, we can expect to find a total of fifty-one positive results, where only one comes from someone with the disease. This leaves a chance of one in fifty-one (or about 2 percent) that you're actually sick when you get a positive test result.

yourself suddenly shuffling off the mortal coil if you fall on your head or get coughed on by the wrong stranger. Or you may find yourself the beneficiary of a surprising medical discovery and live a considerably longer life than you currently have good reason to anticipate. It is prudent to keep in mind that your predictions may be wrong for an individual even when they are right on average. But although your estimated remaining lifetime may be off the mark by decades, you can be quite confident it will not be off the mark by centuries.*

With these estimates at hand, you may want to adopt a certain lifestyle to increase the number of years you likely have left. Exercising more and smoking less might add a few years, though changing habits can be a serious challenge. A number of techniques have been proposed to shift people's priorities towards downstream outcomes—from penning a letter addressed to themselves twenty years from now to interacting with an age-progressed virtual reality rendering of their own face.[49] We saw earlier that vividly imagining the future might be one avenue for offsetting the tendency to discount future consequences, perhaps by allowing you to appreciate the future benefits of your patience more fully in the present. In trying to quit smoking, it may pay to forecast positive alternatives in as much detail as possible, such as regaining your sense of smell, watching your bank balance grow, or sitting in a deck chair by the seaside during your extra retirement years.[50]

David Hume claimed in 1739 that humans are not able "to cure, either in themselves or others, that narrowness of soul, which makes them prefer the present to the remote." Maybe so. But, as we have seen, there are workarounds.[51]

Jerry Seinfeld jokes that, in the evenings, he is "night guy"—he likes to stay up late. But what about being exhausted tomorrow? Well, that's not night guy's problem—that's "morning guy's" problem. Night guy always screws morning guy, and there's nothing that morning guy can do to get back at him. Even internally there can be battles between different time

---

* To account for how far off the mark predictions can be, prudent forecasts today routinely come with estimates of error (the party is predicted to attract 52 percent of the vote, but this figure still has a 3 percent margin of error) and certainty (tomorrow has a 20 percent chance of rain).

perspectives. But Seinfeld is not quite right. There are many things morning guy could do about this dilemma. Anticipating future temptation, morning guy can toss out the booze and ask a friend to check in on night guy's antics. Morning guy could try to innovate his own creative solutions, or he could seek to learn from the solutions others have created and, say, check into a clinic.

Evidently, one workaround is to precommit to a course of action. You can make it costly to retreat from a plan by giving your word in public, staking your reputation on seeing things through. Or you might sign up for a twelve-month gym membership, knowing that the waste of money from nonattendance will serve as an effective motivator. Anticipating your reluctance to rise in the morning in a timely fashion, you can buy alarm clocks that are not just loud but spring from your bedside and drive around the room on wheels, or that require you to complete a puzzle to turn them off. Even simply removing temptations from view can help, which is why many of us keep our cookies and chocolate at the back of the pantry. Of course, even so, we can still fall short and fail to align our behavior with our intentions.[52]

When things do not go to plan, it is often wise to review the case and try to learn from it. We trust that any kind of aviation accident or even near miss, for example, will be reported and forensically assessed. Even when nothing has yet gone wrong, it might pay to conduct what the psychologist Gary Klein calls a "premortem." Say a company wants to launch a new product or a government wants to install an electric vehicle charging network. In a premortem, the team would imagine that the project has failed and then work backward to try and figure out the reasons why. This exercise encourages people to reflect on what could possibly go wrong and air skeptical opinions they otherwise might keep to themselves.[53]

The founders of many nations applied a similar logic when anticipating what might go wrong in the future and committing to a course of action that would prevent those wrongs from manifesting. It is for good reason that many constitutions enshrine the separation of executive, judiciary, and legislative powers, or limit the number of terms a leader shall be allowed to rule. The political scientist Jon Elster has explained how constitutions act as tools for "self-binding." In his book *Ulysses Unbound*, Elster compares the writing of a constitution to the moment in Homer's *Odyssey* when Ulysses has himself tied to the mast of his ship so he cannot act on the deadly temptation of the sirens.[54] In much the same way as Ulysses's ropes, constitutions set up rules in the present so that in the future—when things get heated, emotions boil over, or temptations arise—citizens do not deviate too far from their own best interests. Anticipating regret can help us avoid the real thing.

———

Can we avoid being consumed by our successes?
—Jonas Salk (1992)

Why don't we see evidence of extraterrestrial civilizations, despite the enormity of the cosmos and the multitude of other potentially habitable planets even in our local Milky Way?*[55] One possibility is that all sufficiently advanced civilizations drive themselves into oblivion through their own powers. Humanity may be no exception. Today's generation is one of the first with the capability to not only cause the extinctions of other species but to genuinely imperil our own. Perhaps that is the real reason nobody attended Stephen Hawking's party for time travelers from the future.

Apocalyptic events have been prophesized for many centuries, such as in the Book of Revelation or the writings of Nostradamus.[56] But in recent decades, some forecasters have begun trying to systematically assess the dangers that might do us all in. A global premortem, if you will. Researchers at the University of Oxford's Future of Humanity Institute and at the University of Cambridge's Centre for the Study of Existential Risk, for instance, are attempting to estimate the odds of an existential catastrophe.[57] While

---

* The discrepancy between the enormity of the cosmos and the lack of evidence for alien life is known as the Fermi paradox, named after Enrique Fermi. Fermi, who was one of the scientists working on the Manhattan Project, is said to have blurted out the question "Where is everybody?" over lunch at Los Alamos, New Mexico, where the atomic bomb was being brought into existence.

people disagree on the exact numbers, most agree that we are much more likely to bring about a cataclysm by ourselves—nuclear war, climate change, or bioengineered pandemics—than fall victim to an ultimate natural disaster such as an asteroid impact, supervolcano eruption, or stellar explosion. Whatever the actual probabilities, these calculations serve as a warning that we better pay attention. After all, we can only rely on foresight to reduce and prepare for such dangers, given that if we are blown out of existence, no one will be around to do better next time.[58]

NASA's Center for Near-Earth Object Studies, for instance, has developed a collision monitoring system that scans the skies for potential asteroid impacts in the next one hundred years. And they run regular large-scale simulations testing our planetary defense mechanisms, including how we might deflect the incoming body or evacuate potential disaster zones. In 2021, NASA launched a first attempt to actually alter the path of an asteroid, sending a spacecraft to collide with the pyramid-sized Dimorphos.[59] The dinosaurs never stood a chance before their asteroid hit. Yet, as the philosopher Daniel Dennett put it, "Now, for the first time in its billions of years of history, our planet is protected by far-seeing sentinels, able to anticipate danger from the distant future . . . and devise schemes for doing something about it. The planet has finally grown its own nervous system: us."[60]

How can we encourage more people to take a bigger perspective and care about global threats? Step back for a moment to consider the first photographs ever taken of the whole Earth from space. Here we all are, sitting on this one rock together, our national boundaries rendered arbitrary and meaningless, our whole world so interconnected and fragile. The image of our planet, hanging in the vastness of space, gave many earthlings a new sense of unity, a symbol to rally around. In 1976, Carl Sagan asked NASA to turn around Voyager 1 as it hurtled towards the outer reaches of the solar system so that it could point its camera homewards and take one last photo of our globe. The resulting image of what Sagan called our "pale blue dot" showed our world as a tiny speck. In some sense, we are just as infinitesimally small when it comes to our shared, fleeting sliver of time. Here we all are, some 13.8 billion years after the Big Bang, with eons stretching out ahead. Perhaps we need some symbols to draw attention to this bigger time frame. A collective called the Long Now Foundation, for instance, has been building a giant mechanical clock to tick for the next ten thousand years. Stewart Brand—one of the designers of the clock—hopes that it will "do for thinking about time what the photographs of Earth from space have done for thinking about the environment."[61]

With a common view of a larger time frame, perhaps we may be able to motivate people to more readily commit to longer-term projects. There are precedents of cooperation across multiple generations. Some of the most ambitious endeavors throughout history—building cathedrals, constructing fortifications, the scientific enterprise to accumulate knowledge—have required such collaborative efforts between humans who never shared a day on Earth but nonetheless shared a vision.

But it is tricky to enforce a social contract across generations.[62] Future people might just trash the place regardless. And, in turn, they can't punish us for failing them. Sure, they could kick over our gravestones or bulldoze our monuments, but their options for retaliation are pretty toothless relative to what we can do to them: torching forests and dumping toxic sludge into the oceans, for starters. Still, thanks to foresight, people take untold meaning from the thought of their continuing lineage, of their children and their children's children. Their legacy.

In his book *The Good Ancestor*, the philosopher Roman Krznaric argues that much more needs to be done to ensure the well-being of future generations.* He points out that several countries have begun to implement efforts to enhance intergenerational solidarity. Sweden and Finland have a Minister of the Future and a Committee for the Future, respectively. And Wales

---

* In Japan, the Future Design movement organizes citizen assemblies where some participants role-play as if they are from the future and debate with citizens from the present. In one exercise in Yahaba, residents discussed how to administer the town's water supply. The water utility was highly profitable, so residents from the present pushed for surplus cash to be funneled towards reducing the present price of water. But the residents from 2060 instead presented a plan to actually increase water costs, and to invest or save the surplus towards improving the waterworks facility—which they realized would need an expensive technological and engineering overhaul sometime between the present and the year 2060. Ultimately the town adopted some of the recommendations from the "residents from the future."

appointed a Future Generations Commissioner, charged with advocating for the rights of tomorrow's citizens.[63] But the devil is in the detail. The legislation that established the Welsh role includes the following clause: "A public body must take all reasonable steps to follow the course of action set out in a recommendation made to it by the Commissioner . . . unless . . . it decides on an alternative course of action." To political decision-makers, future problems can pale in comparison to present demands.

And, of course, many long-term problems are highly complex with no easy solutions. Take the question of what to do with nuclear waste, which can remain extremely dangerous for thousands, if not hundreds of thousands, of years.[64] Among the challenges with nuclear waste management is that the discarded material must be stored with appropriate warnings for future people of its whereabouts, but, paradoxically, those warnings also shouldn't be too obvious in case a malicious actor wants to access the material—even many millennia later. How can we communicate over such a vast span of time? There is no guarantee that whoever is on the receiving end of our messages would understand any of our languages. The Organisation for Economic Co-operation and Development's Nuclear Energy Agency has therefore devised an initiative on the Preservation of Records, Knowledge, and Memory Across Generations, in dialogue with fourteen countries with advanced nuclear technology. After extensive deliberation, they determined that there was "no single approach or mechanism that would achieve, on its own," the goal of shepherding critical information about nuclear waste repositories to future generations. Instead, they recommended thirty-five different mechanisms for safely transferring this knowledge, including installing markers, burying time capsules, and maintaining international catalog.*[65] Rather than relying on some single-track vision of the future, the key,

---

* Some of the more fanciful early ideas for signposting the dangers of nuclear deposits include genetically engineering cats so they will turn green when exposed to nuclear radiation (letting our pets act as a kind of canary in the coal mine) or putting spikes all over the ground above a buried disposal site to mark the no-go zone. Then there was the idea to create a secretive cast of nuclear message keepers, a sort of Knights of the Nuclear Order, who would convey the knowledge across generations through initiation rites and encoded rituals.

here as elsewhere, is to be humble about our ability to foresee, to entertain multiple possible solutions, and to weigh up their relative costs and benefits.

Future generations might expect us to have secured more than just our hazardous trash, of course. In her book *The Optimist's Telescope*, climate policy expert Bina Venkataraman suggests the approach of keeping shared heirlooms—common knowledge and resources to be shepherded by each generation to the next. Perhaps it would be important to protect our libraries of hard-won lessons. Recall that with the demise of the Great Library of Alexandria, much ancient knowledge was lost; what would we do if the internet went dark?

Future generations might also want us to guard the viability of our crops. Nestled deep in the permafrost of the Norwegian island of Svalbard, near the North Pole, lies an attempt to do just that. The Global Seed Vault was created to store the seeds of crops in case of catastrophe, preserving the crucial agricultural heritage of humanity for hundreds of years to come. We could invest in storing the seeds of other plants, or even the eggs and sperm of animals that

Figure 8.3. The Svalbard Global Seed Vault.

might go extinct—building some kind of intergenerational Noah's ark so that one day these organisms may be resurrected. Whether this is the best use of our efforts is questionable, given that so many extant species are critically endangered and could benefit from resources for their protection. But in the millennia to come, our descendants may find it invaluable to have the option not just to replant but also to repopulate. What other treasures would be worth carefully securing for the future?

All told, long-term thinking and intergenerational cooperation might seem like unequivocally good ideas for making a better world. But this is not a given.

Consider that when individual people nobly attempt to create a lasting inheritance for their children, they may in the process contribute to escalating inequality by restricting their accumulated wealth to their bloodline. It was Herbert Spencer who said the wise man "must remember that while he is a descendant of the past, he is a parent of the future."[66] But such farsighted visions also inspired eugenics programs aimed at shaping future generations by curtailing the reproductive rights of people deemed unfit to breed. Foresight can be used for reprehensible ends, such as the meticulously planned Holocaust amid Hitler's attempt to lock in a thousand-year Reich. Foresight is ethically neutral—it is an awesome power that can be put to work for great good or for great evil.

Yes, we must learn to harness foresight even more if we are to have any hope of addressing the global longer-term challenges humanity is facing. We need to more effectively predict, plan, and prepare, strategize, simulate, and scientifically assess. But at the end of the day, it is our responsibility to determine what has priority. We may even conclude that our best use of foresight is not to aim at change or growth but at a sustainable equilibrium. While foresight can tell us what could happen given a particular course of action, it is up to us to decide what *should* happen.

Future generations may hold us morally responsible for our choices. Even if we have the best intentions, they may consider us negligent or reckless if we fail to anticipate avoidable danger or fail to act with sufficient determination to protect future prosperity. *Homo sapiens* now finds itself at a critical juncture. Many of the alternative timelines ahead of us look very bleak: climate change, nuclear wars, and bioengineered pandemics are but a few of the self-imposed threats we face. What, if anything, will send us to the brink? If you've been following along, you'll recognize the answer: we don't know. But unless we want to go the way of the dodo, it will pay to cover our bases.

———

We must not give up hope. We have all the tools we need, the thoughts and ideas of billions of remarkable minds and the immeasurable energies of nature to help us in our work. And we have one more thing—an ability, perhaps unique among the living creatures on the planet—to imagine a future and work towards achieving it.
—David Attenborough (2020)

Prediction is at the core of brain function across the animal kingdom. But our foresight, as Attenborough suspects, does indeed set us apart from other animals. Our ancestors gradually acquired their remarkable mental time machines over millions of years, leaving clues of their advancing capacities in the form of carefully crafted stone tools and the remnants of firepits. Ever since the invention of tomorrow, early humans have strived to better foresee and control the future. Recognizing the future utility of solutions and of teaching others, they set in motion a feedback loop of cultural accumulation. Foreseeing what might lie ahead, they assembled stone-tipped spears that would only later kill from a distance and crafted carrying devices that enabled them to ferry provisions to points in space and time, wherever and whenever they might be needed. Humans increasingly set out to acquire skills and knowledge, shaping themselves and their destiny. They noted the regularities of their world and innovated tools like calendars, money, and writing that dramatically improved their ability to coordinate future events. More and more people planted crops that would only be harvested months down the track and constructed artificial worlds designed to satisfy their anticipated wants. The quest to control the future has shaped our world.

Many people today consider the Earth gravely ill and see humanity as a shortsighted scourge. But as we hope to have shown in this book, our species, far from being locked in the present, has come to deal with the future more than any other creature that has ever existed.

It is possible we are already on the right path. Even our biggest ecological challenges may be solved ultimately by farsighted human ingenuity. Maybe we can clean up our mess, rapidly replace all the plastics with biodegradable materials, and swap fossil fuels for renewables. Just as advances in information technology revolutionized our world over the last few decades, so too may advances in biotechnology over the next few decades. Perhaps we can protect diminishing habitats and revitalize endangered animals. We may even bring back species—Tasmanian tigers, moas, and woolly mammoths—that have fallen victim to our successes. It is conceivable that we will be able to sustainably grow what we eat, repair with nanobots what we break, and 3D print whatever else we need without destroying the environment. And while artificial intelligence may pose some existential threat, it may also turn out to be extremely helpful in predicting and preventing catastrophes. Perhaps we can innovate technological solutions to all our problems. Perhaps it is not too far-fetched to think we could even plot our way to world peace.

But then again, perhaps such optimism is misplaced. Our new solutions may beget still greater problems. We will certainly continue to make mistakes, from minor deviations that linger within a thin margin of error to gross miscalculations that spell disaster. It is possible that positive trajectories will not continue, wish as we might, and no one will invent the critical transformative technologies we need in time. Or perhaps we will fail to recognize the future utility of strategies and technologies fast enough even if someone does figure them out—or the lure of short-term profit, let alone the distraction of political conflicts, will stop us from implementing the requisite changes. What's more, optimism may harbor the risk of fostering complacency. Why bring an umbrella when you're sure it won't rain?

Still, optimism—at least of the kind that involves envisioning positive possibilities even in dark times—may also shield us from fatalism and motivate us to actively create a better future. So even when we know that optimism is unlikely to be entirely warranted, we may nonetheless benefit from embracing it. By sharing our knowledge and optimistic predictions, we can encourage collaboration and drive positive change. Just as a placebo effect is worth harnessing, it may pay to keep those rose-tinted glasses nearby.[67] With the right balance between humility and confidence, we can tackle the challenges ahead.

In addition to our remarkable foresight, humans also have a distinctive desire and capacity to wire our minds together with those around us.[68] We ask questions and give advice, helping us foresee the future with more accuracy and devise more fruitful plans. Our urge to connect and our capacity to imagine scenarios also combine when we tell each other stories, including those about tomorrow and beyond. From telling tales around the campfire to modern literature, stories direct our mental time machines on guided tours of events we did not witness ourselves. Stories often contain lessons for the future—the struggles of others can teach us what to pursue, what to avoid, and how to overcome obstacles. Narratives about dystopias—say Orwell's *1984* or Huxley's *Brave New World*—let alone stories about existential catastrophes like alien invasions and robot uprisings, can serve as warnings. And just as we can be motivated to avoid dystopian visions, we can be enticed into action by ideas of positive transformations and better worlds.[69]

We can use our mental time machines to entertain distant scenarios, even of worlds millions of years ahead, when the last traces of our species can be found only in the fossil record alongside all the other hominins, or when we would no longer recognize our own descendants if they did prevail.[70] Our current challenges may appear insignificant when we consider the ultimate fate of our species, our planet, our solar system, our galaxy—or even of the entire cosmos. Will the universe keep expanding and expanding, eventually leading to ever colder temperatures, until no life can be sustained? Or will it at some point start to contract and eventually collapse into a Big Crunch? This could be the end of everything, or—perhaps with another Big Bang— it could be the beginning of a new universe as the process bounces back into being.[71] None of this is known for sure, and none of us will be there to witness what actually happens. So, in some sense, it may feel little different from discussing entirely fictional scenarios such as a better ending to the last movie you watched.

Mostly, our mental time machines carry us on much shorter trips within our slice of time, to memories of events just past and opportunities around the corner. They set us on quests for understanding, for meaning, and for control of our own destiny. They bring us the joys of anticipation and the depths of dread. They make us yearn for days gone by, dream of better tomorrows, toil with purpose, and vow to love until death do us part. They give us our nostalgia and rumination, plots and schemes, worries and trepidations, promises and obligations, our faith, hope, and ambitions—the worst and the best of our intentions. With our mental time machines, we weave the story of our lives, where we come from and where we are going.

Other animals, like humans, sometimes greet each other when they meet. Chimpanzees often grunt to say *hello* and may even give each other a hug and a kiss. But as Jane Goodall points out, they never say *goodbye*. In fact, humans may be the only animals who bid one another farewell in mutual recognition that we are going our separate ways, and often in anticipation that our paths may cross again tomorrow.

We hope you have enjoyed this journey—see you later.

ACKNOWLEDGMENTS

In many ways human foresight seems pathetic. People persistently predict things that do not happen and frequently fail to foresee those that do. We sure did not anticipate we would be interrupted by a global pandemic when we started writing this book in late 2018. Flaws of foresight were on display during the COVID-19 outbreak, as confident forecasts fell widely off the mark and many plans did not quite work out as hoped. But the crisis has also demonstrated how powerful human foresight can be. There were prevention and containment strategies, detailed projections of the spread of the virus, and vaccines rapidly developed to immunize vast populations. Humans can be both misled and empowered by their mental time machines.

We have long been fascinated by the nature of foresight—where it comes from and how it works. In fact, Thomas wrote his master's thesis on mental time travel some thirty years ago and has studied aspects of the topic ever since. Both Jon and Adam did their PhDs on mental time travel under his supervision, and so we thought it was fitting for us to team up to bring the story of human foresight to a wider audience.

While having three authors was very helpful in that we could combine our knowledge and skills (not to mention correct each other's errors and biases), we also drew on the expertise of other people. For their thoughtful comments on draft chapters we would like to thank: Alex Taylor, Andrew Whiten, Bina Venkataraman, Ceri Shipton, Chris Dudgeon, Christopher Krupenye, Claire O'Callaghan, Conrad Leonard, Cristina Atance, Daniel Schacter, Freya Young, Geoff Bonning, Gillian Pepper, James Steeves, John

Sutton, Jon Clindaniel, Lachlan Brown, Manfred Suddendorf, Matt Mc-
Farlane, Michael Corballis, Michelle Langley, Muireann Irish, Niki Harré,
Rachel Mackenzie, Ruben Laukkonen, and Shalini Gautam. Other experts
were kind enough to check sections: Denise Schmandt-Besserat, Hugh Possingham, Magdalena Zych, Peter Evans, and Paul Seli. Some were even
game enough to provide feedback on an entire draft of the book. Thank you
very much to Beyon Miloyan, Bill von Hippel, Brendan Zietsch, and David
Bulley. All the same, any remaining mistakes are of course entirely our own.

Many of our studies have been conducted in collaboration with other
researchers in Australia, Germany, New Zealand, the UK, the USA, and
elsewhere. Science is a collaborative enterprise, and we could not have done
this without their help. Whenever we have referred to "our" studies or "our"
research group, the reference list includes the names of those who were involved (and sometimes that includes only one of us). We are grateful for
essential research funding provided by the Australian Research Council and
National Health and Medical Research Council. A special thanks goes to all
the diverse participants in our studies over the years. Research with children
has been conducted mainly at the Early Cognitive Development Centre and
at the Queensland Museum. We thank the parents and children who do-
nated their time to participate, as well as the ECDC and museum staff for
their ongoing support. Our animal studies have been conducted at zoos,
including in Rockhampton, Perth, and Adelaide, and at Wildlife HQ in
Nambour. No animals were harmed in the process, and they were always
free to stop playing with us. The chimpanzees at Rockhampton Zoo, especially Cassie and Holly, have been a particular joy to work with over the
years. We are thankful for the continuing support of all the staff at these
zoos who are working hard to ensure the welfare of the animals.

We are lucky to have enjoyed a collegial working environment at the University of Queensland School of Psychology. We are especially grateful for the
regular meetings of the Centre for Psychology and Evolution and the Early
Cognitive Development Centre. For support, advice, and productive conversations, we would like to thank our friends, colleagues, and collaborators, alongside those already named above, including Aisling Mulvihill, Andy Dong,
Ashley Hay, Ashley Phelan, Brendan Bo O'Connor, Brian Leahy, Chelsea

Boccagno, Cheryl Dissanayake, Chris Moore, Claire Fletcher-Flinn, Colin Conwell, Daniela Palombo, Derek Arnold, Donna Rose Addis, Fiona Barlow, Frankie Fong, Gabrielle Simcock, Gail Robinson, Gill Terrett, Graeme Strachan, Hallgeir Sjåstad, Hannah Biddell, Ilana Mushin, Isaac Baker, James Sherlock, Jason Mattingley, Jeremy Nash, Jo-Maree Ceccato, Johannes Mahr, John Mclean, Jonathon Crystal, Jordana Wynn, Julie Henry, Justin Williams, Kana Imuta, Karolina Lempert, Kelly Kirkland, Keyan Tomaselli, Kristyn Sommer, Mark Nielsen, Markus Boeckle, Markus Werning, Matti Wilks, Nancy Pachana, Nicola Clayton, Nicole Nelson, Nicholas Mulcahy, Ottmar Lipp, Patricia Ganea, Paul Dux, Paul Jackson, Peter Rendell, Rebekah Collins, Ross Cunnington, Russell Gray, Sam Gilbert, Sally Clark, Sen Chen, Siobhan Kennedy-Constantini, Steven Pinker, Sue Tonga, Susan Carey, Tian Po Oei, Virginia Slaughter, and Young Ji Tuen.

We also had the pleasure of working with many wonderful students on projects relevant to this book, including Alicia Jones, Amanda Lyons, Amy Burgess, Andrew Hill, Ania Ganowicz, Cate MacColl, Chantal Li, Charlotte Casey, David Butler, Deanna Varley, Elizabeth Cullen, Emma Collier-Baker, Giang Nguyen, Jacky Ni, Jacqueline Davis, Janie Busby Grant, Janine Oostenbroek, Jennie Morrissey, Jess Crimston, Jessica Bushell, Jo Davis, Johanna Vandersee, Karri Neldner, Kate Watson, Kelly Brooks, Kristy Armitage, Lily Dicken, Luca Buzac, Maddison Bogaart, Melissa Brinums, Olivia Demichelis, Phoebe Pincus, Rebecca von Gehlen, Regan Gallagher, Sam Pearson, Sarah Batholomai, Talia Leamy, Tegan MacKenzie, Thomas McCarthy, and Zoe Ockerby.

Each of us has enjoyed support from a wide range of people during the writing of this book.

Jon would especially like to thank his parents, Wendy and Stewart, for their generous love and encouragement over the years. He would also like to deeply thank his siblings, Daniel, Penny, and Anna, and his grandma Muriel, for the support, laughs, and happy memories. He is sincerely grateful to his friends at the University of Queensland and beyond, some of whom are listed above but many of whom are not, who have shared in the ups and downs of academic life and have been the source of so much riveting conversation and caring support when needed. Finally, he would like to give a special thanks to his cherished lifelong friends from his hometown of Ipswich, Queensland, especially Ashley Cooper and Christopher Ainslie, who helped him become who he is today.

Adam would like to thank his mother, Louise, for her enduring love and support, and his father, David, to whom this book is dedicated, for caring so deeply and for believing in this project. A special thanks goes to Richard Bulley for his steadfast brotherhood. To Freya Young, thank you for your patience, your optimism, your trust, and your love. A sincere thank-you to all the extended Bulley, Fahey, and Young families, as well as to all the people Adam has been lucky enough to call his friends through the course of this project. Of friends not already listed above, a particular thanks to Adel Djellouli, Alex Lister, Allegra Young, Caroline Burnett and Ajay Patel, Chye-Ling Huang, Daniele Foresti, Evan Roman, Georgia Coonan, the Goodman family, Jacqueline Hatch, Jason Mackenzie-Reur, Jordan Tyldesley, Kai Milliken, Kathryn Kirchner, Lee Geers, Madeleine Coonan, Matt Grant, Matt Hawkins, Matt Michalak, Patrick Tunney, Patrick Darling, Rees Garmann, Sophie Clark and the Clark family, Thomas Cochard, Till Almer, and Trent Munns.

Adam would like to acknowledge the support of the Schacter Memory Lab and the broader community of the Harvard Department of Psychology. He is also grateful for the support of those at the Brain and Mind Centre and School of Psychology at the University of Sydney, especially the Memory and Imagination in Neurological Disorders group led by Muireann Irish.

Thomas would like to thank his spouse, Chris Dudgeon, who is the real biologist in his family, studying the ecology and genetics of sharks, rays, and other amazing animals. Thank you for your passion, unwavering support, and love. A special thanks goes to his kids, Timo and Nina, for participating in endless studies and bringing so much joy. Thanks also to his siblings Ingrid, Manfred, Bettina; their families; and all his friends along the way.

One person in particular has to be singled out: Thomas dedicates the book to his mentor Michael C. Corballis, who sadly passed away before it was finished. He was a towering figure in cognitive neuroscience, a remarkable scholar, a wonderful teacher, an inspiring colleague, and a friend. Without him this book would not exist. When Thomas moved from Germany to New Zealand in 1992 to do his master's degree at Waikato University, he could not find a supervisor for the thesis he wanted to write on the nature of our ability to think about the future. So he approached Mike, at the University of Auckland, who had just written a book on the evolution of brain asymmetry and the generative mind. Mike generously agreed to help Thomas turn his half-baked ideas into a coherent argument.

The thesis proposed that our capacities to remember past events and to imagine future situations are intimately linked in mind and brain, and that the emergence of "mental time travel" in our forebears had been a prime mover in human evolution. This kind of idea was unfashionable at the time, as cognitive psychologists were more interested in memory than foresight, and evolutionary psychologists were focused on domain-specific (modular) adaptations in the brain rather than on such domain-general capacities. After the initial attempts at publishing the thesis failed, Thomas would have thrown in the towel if it hadn't been for Mike's calm, gentlemanly guidance preaching persistence and the growing of a thick skin. Eventually, in 1997, the paper appeared in a low-impact journal. But people began taking note. When Thomas and Mike reviewed progress in an article ten years later, they found the field abuzz with excitement. Mental time travel was being explored in research on animal cognition, in the neurosciences, and in diverse subdisciplines of psychology, including clinical, developmental, and social psychology. *Science* magazine even selected new evidence for links between memory and foresight as one of the top ten scientific breakthroughs of that year.

In 2011, the editor of *Trends in Cognitive Sciences* surprised Thomas with the revelation that Mike had "changed his mind," requesting comment. Mike had written an opinion piece in which he argued that neuroscientific discoveries involving single-cell recordings from the hippocampus indicated that rodents may have something quite similar to human mental time travel. The subsequent published debate was portrayed in *Science* magazine as an "intellectual rift," though Mike and Thomas stayed friends, and both understood that examining what humans have in common with other animals and what sets humans apart are really two sides of the same coin. They agreed that humans do share many fundamental building blocks of our minds with other animals, and that there are some aspects, such as our desire and capacity to exchange our plans and memories, that appear to have no parallel in other species. Thomas and Mike even briefly plotted to hide away in Italy for a while to write a book about all this. But after that plan fell through, Thomas paired up with two of his own students instead to write the book you are holding in your hands. Mike was excited about it and helpful as always, and Thomas remains eternally grateful—and misses him dearly.

The Invention of Tomorrow was eventually pitched to publishers with the help of our dedicated agents, Peter Tallack at the Science Factory and Louisa Pritchard at LPA. We are grateful for their support. Thanks also to all the people who generously provided the images that illustrate the book. Finally, we thank our brilliant editor, T. J. Kelleher, and the outstanding team at Basic Books, including Brittany Smail, Jessica Breen, Kara Ojebuoboh, Lara Heimert, Liz Wetzel, Madeline Lee, Meghan Brophy, Roger Labrie, and Shena Redmond, for helping us turn our plans for the book into reality.

Writing a book about mental time travel took a long time. We are grateful to have shared this journey and already looking forward to fondly looking back at it.

FIGURE CREDITS

Chapter 1. Your Private Time Machine

Figure 1.1: Exterior photo courtesy of Flickr user Tilemahos Efthimiadis, CC BY 2.0, recreation courtesy of Freeth et al., 2021, Scientific Reports, CC BY 4.0

Figure 1.2: Courtesy of the authors

Figure 1.3: Stephen Hawking's Time Travellers Invitation—Peter Dean / kiteprint.com

Chapter 2. Creating the Future

Figure 2.1: Image in public domain

Figure 2.2: Courtesy of Musée du Louvre, Départment des Antiquités Orientales and Dr. Schmandt-Besserat

Chapter 3. Invent Yourself

Figure 3.1: Courtesy of the authors

Figure 3.2: Courtesy of the authors, thanks to Remy

Chapter 4. Under the Hood

Figure 4.1: Courtesy of Fatma Deniz

Figure 4.2: Courtesy of Flikr user Alex Pepperhill, CC BY 2.0

Chapter 5. Are Other Animals Stuck in the Present?

Figure 5.1: Charles Eden Wellings, between 1900 and 1922, *Whaleboat fast to a whale, Twofold Bay*. National Library of Australia, PIC/8539/55

Figure 5.2: Pixabay / Mark Thomas

Figure 5.3: Courtesy of Cameryn Brock

Figure 5.4: Courtesy of Yi-Ru Cheng

Figure 5.5: Courtesy of the authors

Figure 5.6: Courtesy of Yvette Fenning

Figure 5.7: Courtesy of Patrick Wood

Chapter 6. Discovery of the Fourth Dimension

Figure 6.1: Courtesy of the authors

Figure 6.2: Reconstruction by W. Schnaubelt and N. Kieser. Photo by Régis BOSSU/Sygma via Getty Images

Figure 6.3: Photos by Sally Clark, stone tools courtesy of Ceri Shipton

Figure 6.4: Courtesy of the Regional Archaeological Museum, Alcalá de Henares, Madrid, Spain. Photo by Cristina Arias/Cover/Getty Images

Chapter 7. Travel Tools

Figure 7.1: Courtesy of Landesamt für Denkmalpflege und Archäologie Sachsen-Anhalt, Gert Pie

Figure 7.2: Courtesy of Landesamt für Denkmalpflege und Archäologie Sachsen-Anhalt, Juraj Lipták

Figure 7.3: Courtesy of the British Museum

Figure 7.4: Courtesy of the Saxon State and University Library Dresden (SLUB). Copyright SLUB Dresden / Digital Collections / Mscr.Dresd.R.310

Figure 7.5: Courtesy of Jon Clindaniel

Figure 7.6: Courtesy of the Yap Visitors Bureau

Chapter 8. Our Slice of Time

Figure 8.1: Courtesy of the Science Museum Group

Figure 8.2: Courtesy of the authors

Figure 8.3: Courtesy of Wikimedia users Subiet / Hunster, CC BY-SA 4.0

NOTES

Chapter 1. Your Private Time Machine

1. For details about Ötzi's final days and his equipment see Capasso, 1998; O'Sullivan et al., 2016; Oeggl et al., 2007; Peintner et al., 1998; Wierer et al., 2018. Note that the bow was unfinished, as were arrows in his quiver. Ötzi's story has been brought to life in Felix Randau's 2017 movie *Iceman*.

2. Suddendorf & Corballis, 1997.

3. Oliver, 1993.

4. There are many examples of such biases in planning (Buehler et al., 2010; Kahneman, 2011) and expectation (Hanoch et al., 2019; Sharot et al., 2011).

5. Freeth et al., 2021; Seiradakis & Edmunds, 2018.

6. Cicero, 45 BC/1877.

7. Kilgannon, 2020; Kolb, 2019.

8. Safire, 1969.

9. NASA, 2003.

10. Redshaw et al., 2018; Redshaw & Suddendorf, 2016. Our research group is not a fixed entity, as the three of us collaborate with a variety of researchers at different institutions and across different disciplines. At the time of writing, Thomas and Jon were based at the University of Queensland and Adam was on a fellowship at the University of Sydney and at Harvard University. When we refer to "our" work, it sometimes involves all three of us, but often it is only one or two of us and a host of other wonderful collaborators (as evident in the supporting references).

11. Redshaw & Suddendorf, 2020.

12. Gautam et al., 2019.

13. Darwin continues: "When thus reflecting I feel compelled to look to a First Cause having an intelligent mind in some degree analogous to that of man; and I deserve to be called a Theist. This conclusion was strong in my mind about the time, as far as I can remember, when I wrote the *Origin of Species*; and it is since that time that it has very gradually with many fluctuations become weaker. But then arises the doubt—can the

mind of man, which has, as I fully believe, been developed from a mind as low as that possessed by the lowest animal, be trusted when it draws such grand conclusions?" (Darwin, 1887/1958).

14. Holden, 2005. The "time machine" became a popular notion with H. G. Wells's (1895) novel of the same name, opening up a whole genre of science fiction in literature and film.

15. Episodic memory: Tulving, 2005. Episodic foresight: Suddendorf & Moore, 2011.

16. Wearing, 2005.

17. Schacter, 1999.

18. Hoeffel, 2005. This terrible injustice is probably not as unusual as one would like to think. The proportion of false convictions among people on death row in the United States is conservatively estimated to be around 4 percent (Gross et al., 2014).

19. Bartlett, 1932; Loftus, 2001.

20. Research on rodents shows that memory storage can be blocked or boosted with certain drugs. The neuroscientist Karim Nader and others have found similar effects when drugs are administered right after retrieval of an old memory. This suggests that when memories are retrieved, they are vulnerable to being updated, distorted, and otherwise altered. Despite some popular beliefs, there may be no opportunity to somehow rediscover the originally stored memory, be that through some special therapy, drugs, or hypnosis (Lee et al., 2017).

21. Suddendorf & Busby, 2005.

22. Eustache et al., 1996.

23. Klein et al., 2009.

24. Suddendorf & Corballis, 2007.

25. Commonalities between memory and foresight: in children (Busby Grant & Suddendorf, 2005; Suddendorf, 2010), aging (Addis et al., 2008), clinical conditions (D'Argembeau et al., 2008; Miloyan et al., 2014), and phenomenology (D'Argembeau & Van der Linden, 2004; Trope & Liberman, 2010). Theoretical perspectives on these commonalities: Dudai & Carruthers, 2005; Schacter et al., 2007, 2012; Suddendorf & Corballis, 1997, 2007.

26. Suddendorf & Redshaw, 2013.

27. Suddendorf & Corballis, 2007.

28. Suddendorf, 2006.

29. In academic writing we call this *metaforesight*. The word *meta* comes from the Greek, meaning "higher" or "beyond", so here it refers to thinking about foresight itself (Redshaw, Bulley, et al., 2019; Bulley, Redshaw, et al., 2020). Consider that we may choose to get a painful dental surgery over and done with quickly because we realize we would otherwise dread the appointment. Conversely, we may choose to postpone something delightful because we look forward to looking forward to it (Loewenstein, 1987). In other words, we anticipate our own anticipation.

30. Tinbergen, 1963.

31. NASA's Voyager 1 probe made history in 2012 as the first human-made object to enter the interstellar medium—the cold, dark space between stars (Gurnett et al., 2013).

32. Lewis & Maslin, 2015; Waters et al., 2016.

33. Nielsen et al., 2016.

Chapter 2. Creating the Future

1. For the quote by Alan Kay, see Hiltzik, 1999. For the letter by Fanny Burney, see Holmes, 2008.

2. Davy, 1800.

3. In his book *The Age of Wonder*, Richard Holmes relays Burney's vivid account of the mastectomy to illustrate the enormous human costs of the delay in bringing anesthetics to the bedside (Holmes, 2008).

4. Green, 2016; Gillman, 2019; Haridas, 2013. Wells's first attempt to convince the medical establishment did not go well, but the cat was out of the bag. Chloroform and ether were quickly deployed to similar effects as the use of anesthesia spread across the globe.

5. Tooby & DeVore, 1987.

6. The loss of species associated with human migrations across the planet is known as the late Quaternary extinction event. There has been ongoing debate about the relative contributions of human activity and natural climate change during this period (Koch & Barnosky, 2006), with considerable work suggesting that humans were a primary driver of extinctions (Sandom et al., 2014; Smith et al., 2018).

7. Henrich, 2015.

8. Boyd et al., 2011; Heyes, 2018; Mesoudi et al., 2006; Sterelny, 2012; Tennie et al., 2009.

9. As Henrich (2015) put it himself (pp. 5–6), "The striking technologies that characterize our species, from the kayaks and compound bows used by hunter-gatherers to the antibiotics and airplanes of the modern world, emerge not from singular geniuses but from the flow and recombination of ideas, practices, lucky errors, and chance insights among interconnected minds and across generations."

10. The first printed copy of Charles Darwin's autobiography was edited by his son Francis, who removed potentially offensive musings about religion. Later, a version was published with edits by Darwin's granddaughter, geneticist Nora Barlow, who restored the omissions (Darwin, 1887/1958).

11. Dawkins, 1986.

12. Meltzoff & Moore, 1977, 1983; Meltzoff & Decety, 2003.

13. Oostenbroek et al., 2016.

14. Keven & Akins, 2017; Oostenbroek et al., 2019; Redshaw, 2019; Redshaw et al., 2020. In a comprehensive meta-analysis, our research group assessed whether differences in the methods used across studies over the last forty years might account for differences in findings. The only factor that predicted whether infants were reported to have copied adults was which research group conducted the study (Davis et al., 2021).

15. Heyes, 2016. Heyes's article was titled "Imitation: Not in Our Genes." However, we are not so sure that imitation is "not in our genes" because not everything "in our genes" must be present at birth; just think about puberty.

16. Whiten et al., 1999.

17. For a review of the evidence on nonhuman animal culture, see Whiten, 2021.

18. Mercader et al., 2007.

19. Schuppli & van Schaik, 2019.

20. Horner & Whiten, 2005.

21. Derex et al., 2019. There were two factors that affected how quickly the wheel would roll down the slope: its center of mass (whether it had more weight around the top spoke,

which would get it rolling better at the start) and its inertia (how mass was distributed around the turning axis).

22. Kendal, 2019; Osiurak et al., 2021.

23. The role of foresight in cultural evolution has been a topic of considerable debate since cultural evolutionary theories were first proposed. For an overview, see Mesoudi, 2008.

24. See arguments by Vale et al. (2012) for the role of foresight in cultural evolution. See also Osiurak & Reynaud, 2020; Mesoudi, 2021.

25. Thornton & McAuliffe, 2006.

26. Caro & Hauser, 1992; Hoppitt et al., 2008. For a long time, it was thought that our closest animal relatives, chimpanzees, did not even appear to show teaching defined in that functional way (Hoppitt et al., 2008). But more recent evidence suggests that chimpanzees occasionally provide some help when others learn to crack open nuts (Boesch et al., 2019) and fish for termites (Musgrave et al., 2020). Even if one grants these examples as instances of purposeful teaching, the number of activities that could feasibly be called teaching remains low. Go to any preschool or playground and you'll observe a wider range of purposeful teaching within a few minutes.

27. Lepre et al., 2011.

28. Pargeter et al., 2019.

29. Morgan et al., 2015. In one experiment, researchers arranged novice participants into chains, like in the wheel experiment described earlier. When experienced toolmakers could show the students slowly how to do it or mold their grasp, the quality and transmission of the students' subsequent tool production was improved. Unsurprisingly, when experienced toolmakers were allowed to also explain the manufacturing process verbally, the quality and transmission of production was better still (Balter, 2015).

30. Dean et al., 2012.

31. Lake Eacham: Nunn et al., 2019. For more on the use of mental time travel to inform others, see Mahr & Csibra, 2018.

32. Weir et al., 2002. Research by psychologist Sarah Beck and her colleagues has documented how children, when not given the opportunity to imitate others, develop the capacity to spontaneously make hooks to solve the bucket task (Beck et al., 2011).

33. al-Jazari, 1206/1974. Psychologists traditionally measure creativity with "divergent thinking" tasks, for instance by asking people to come up with as many alternative uses as possible for simple household objects, and then scoring the number of ideas and the novelty of each one. What uses can you think of for a brick? An unoriginal answer might be using it to build a house, while a more creative one might be to bundle it in Christmas wrapping and give it to a nemesis as a gift (Guilford, 1967). Some research has linked divergent thinking to brain networks that are also responsible for mental time travel, supporting the notion that creativity is, fundamentally, about recombining ideas from memory to come up with something new (Beaty et al., 2019). Today, with ready-made products and a little help from our friends, we can overcome most everyday problems, meaning relatively few people even feel a need to innovate technical solutions. Still, some 3 to 6 percent of people in large samples from Finland, Japan, the UK, and the US report having created something new or modifying an existing product in the last few years (De Jong et al., 2015; von Hippel et al., 2012).

34. von Hippel & Suddendorf, 2018.

35. Though note that there could be various ecological reasons for not innovating wheel-based devices, such as a lack of flat surfaces to traverse, no beasts of burden to pull wheeled vehicles, or current technologies that suffice for the desired jobs (Bulliet, 2016).

36. Hero of Alexandria, ca. 62 AD/1851.

37. Diggins, 1999.

38. Ward, 2014.

39. Pinker, 2006.

40. For discussions of the idea that humans inhabit a social, cultural, and/or cognitive niche, see Pinker, 2010; Sterelny, 2007; Whiten, 1999; Whiten & Erdal, 2012.

41. Schmandt-Besserat, 1981.

42. Kramer, 1949.

43. Newton made this famed remark in a letter to Robert Hooke in 1675. See Gleick, 2004.

Chapter 3. Invent Yourself

1. Suddendorf & Corballis, 2007; Suddendorf & Redshaw, 2013.

2. Nielsen & Dissanayake, 2004.

3. Lillard, 2017.

4. Baddeley, 1992.

5. A large study documented a linear increase in working-memory capacity between age four and eleven (Alloway et al., 2006). In line with that growth of their inner stage, children become increasingly capable of entertaining complex narratives, dependent on multiple chunks of information and the relationships between them (Balter, 2010; Halford et al., 1998).

6. Slaughter & Boh, 2001.

7. Research has found that chimpanzees can also solve this task (Collier-Baker & Suddendorf, 2006).

8. Suddendorf & Whiten, 2001.

9. For details about Einstein, see Kaku, 2021. Spelke and Kinzler (2007) propose that infants have a suite of "core knowledge" about objects, actions, numbers, and space, but causal reasoning continues to develop during the toddler (Walker & Gopnik, 2014), preschool (Kuhn, 2012), and early childhood years (McCormack et al., 2018). Children show compelling signs of deductive reasoning by age five. Again, this can be shown with studies involving the hiding of stickers in cups. In one task, we present children with two distinct pairs of cups (A & B and C & D) and hide a sticker in one cup of each pair (A or B and C or D). When we show them that one cup in a pair is empty (say A), it makes sense to search in the other cup of that pair (namely B) rather than to take their chances with one of the other pair (C or D). However, when we visibly remove a sticker from one cup (i.e., A was correct but is no longer an option), then picking one of the other pair of cups (C or D) is the best option, as they still have a 50 percent chance of success. By age five, children consistently make the best choices in both versions of the task. They can evidently use deductive reasoning to optimize their chances of success (Gautam et al., 2021; Mody & Carey, 2016).

10. Piaget & Inhelder, 1958.

11. Intentional stance: Gergeley et al., 1995; Liu et al., 2019. Dennett defines the intentional stance as "the strategy of interpreting the behavior of an entity (person, animal,

artifact, whatever) by treating it as if it were a rational agent who governed its 'choice' of 'action' by a 'consideration' of its 'beliefs' and 'desires'" (Dennett, 2009). False belief understanding: Wellman et al., 2001; Wimmer & Perner, 1983. Development of *theory of mind*: Wellman et al., 2011; Wellman & Liu, 2004. Development of lying: Bigelow & Dugas, 2009. Understanding of complex social interplays: Baron-Cohen et al., 1999; Weimer et al., 2012.

12. Owens, 2008. See also Wells, 1985.

13. Redshaw & Suddendorf, 2016; Redshaw, Suddendorf, et al., 2019. For earlier studies using conceptually similar methods, see Beck et al., 2006; Robinson et al., 2006. For further discussion of children's performance on the forked-tube task, see Leahy & Carey, 2020; Redshaw & Suddendorf, 2020.

14. Rafetseder et al., 2010; Rafetseder & Perner, 2014. But see Nyhout & Ganea, 2019.

15. Zelazo, 2006.

16. Development of executive control: Diamond & Taylor, 1996. Infants begin to withhold some automatic responses as early as the end of the first year of life, but only by around age four do they tend to follow rules even when the rules contradict something they have learned earlier. In general, the development of executive control appears to include three major transitions: from perseveration to responding based on external signals, from reactive control to proactive control, and from environmental triggers to autocuing (Davidson et al., 2006). Improvement into adolescence: Luna et al., 2004.

17. Developmental psychology has a long history of studying children's memory (Bauer, 2007; Fivush, 2011).

18. Parents reported that 40 percent of three-year-olds, 58 percent of four-year-olds, and 70 percent of five-year-olds understood the word *tomorrow*, whereas the percentages for understanding *next week* were 21 percent, 33 percent and 54 percent, respectively (Busby Grant & Suddendorf, 2011).

19. Parent-child conversation has been linked to children's memory (Fivush et al., 2006) and future time concepts (Hudson, 2006).

20. Busby & Suddendorf, 2005; Suddendorf, 2010.

21. Lyon & Flavell, 1994.

22. Busby & Suddendorf, 2010.

23. With Michael Corballis, Thomas proposed the following strict behavioral criteria to rule out potential alternative explanations: "(1) Use of single trials to avoid repeated exposure to the same stimulus-responses relationships; (2) Use of novel problems to avoid relevant learning histories; (3) Use of different temporal/spatial contexts for the critical future-directed action to avoid cuing; (4) Use of problems from different domains to avoid specific behavioral predispositions" (Suddendorf & Corballis, 2010).

24. Redshaw & Suddendorf, 2013; Suddendorf & Busby, 2005; Suddendorf et al., 2011.

25. Atance, 2015; Suddendorf, 2017.

26. Bulley, McCarthy, et al., 2020. It wasn't a permission issue either: we told the kids several times they could draw on any of the cups whenever they liked, and even encouraged them to do so.

27. Children of this age also begin to use external strategies in other memory tasks (Armitage et al., 2022) and when they manually rotate upside-down images instead of mentally rotating them (Armitage et al., 2020).

28. Busby Grant & Suddendorf, 2009; Friedman, 1990. For further discussion of temporal reasoning in children and animals, see Hoerl & McCormack, 2019; Martin-Ordas, 2020.

29. Atance, 2015; Atance & Meltzoff, 2005; Bélanger et al., 2014; Caza et al., 2021.

30. Ghetti & Coughlin, 2018; Lagattuta, 2014.

31. Zimbardo & Boyd, 1999.

32. Košťál et al., 2015.

33. Kooij et al., 2018.

34. Casey et al., 2011; Mischel et al., 1989; Mischel et al., 2011. A more recent study found the effects of the original marshmallow studies were not as strong when accounting for socioeconomic status and other cognitive variables (Watts et al., 2018), but for a critical take on this replication effort, see Doebel et al., 2019; Falk et al., 2020.

35. Kidd et al., 2013. Mischel had himself made similar observations decades earlier. In 1974, he wrote: "A person's willingness to defer immediate gratification depends to a considerable degree on the outcomes that he expects from his choice. Of particular importance are the individual's expectations that future delayed rewards for which he would have to work and/or wait would actually materialize, and their relative value for him. Such expectations or feelings of trust depend, in turn, on the person's history of prior promise-keeping."

36. This was the case even after controlling for the GDP per capita in each country. Of course, correlations like this one are hard to interpret—the causality is opaque and there could be other relevant variables involved that we didn't measure (Bulley & Pepper, 2017).

37. Brezina et al., 2009.

38. Augenblick et al., 2016.

39. Bulley et al., 2016; Bulley & Schacter, 2020; Lee & Carlson, 2015.

40. Bulley et al., 2017.

41. Drabble, 2015.

42. Records retrieved from guinnessworldrecords.com and accurate as of July 2021.

43. Polden, 2015; Villar, 2012.

44. Ericsson et al., 1993; Gladwell, 2008.

45. Macnamara et al., 2014; Macnamara & Maitra, 2019.

46. Macnamara et al., 2016.

47. See the American Heart Association (cpr.heart.org) CPR recommendations, retrieved December 2021.

48. Suddendorf et al., 2016.

49. Brinums et al., 2018; Davis et al., 2016.

50. We have since replicated this result and found that children from age six onwards show explicit understanding of practice and engage in it without being prompted (Brinums et al., 2018).

51. Suddendorf et al., 2016.

52. Biederman & Vessel, 2006.

53. Brinums et al., 2021.

54. One thing teachers seek out is information on how children learn. This allows people to plan how to become more accomplished in training someone else. Research suggests, for instance, that practicing across multiple short sessions tends to be more effective than massed practice (Donovan & Radosevich, 1999). Breaks between sessions provide the brain

with more opportunities to consolidate learning, and, interestingly, similar benefits appear to come from taking a well-earned (and well-timed) nap (Mazza et al., 2016).

55. Miloyan & Suddendorf, 2015.

56. Coffman, 1990.

57. Pham & Taylor, 1999.

58. Oettingen & Mayer, 2002; Oettingen & Reininger, 2016.

59. Some advice now suggests inserting an S between DR and ABC—indicating you should Send for help. And sometimes a D is also added to the end, to remind people of the final option of Defibrillation.

60. Repeated testing: Butler, 2010; Larsen et al., 2009; Chunking: Gobet et al., 2001. Staying alive: Hafner et al., 2012.

61. Sharpe, 2019.

62. And human dog breeders have selected for comparable predispositions that make such training easier (Suddendorf et al., 2016).

63. Biron, 2019.

64. Maslow, 1943; Wahba & Bridwell, 1976.

65. Suddendorf & Redshaw, 2013. There are of course variations in developmental trajectories, and some neurodevelopmental disorders that are associated with impairments in foresight. For a meta-analysis of mental time travel in autism spectrum disorders, for instance, see Ye et al., 2021.

66. From Sartre's lecture "Existentialism Is a Humanism," given on October 29, 1945, in Paris.

67. Gautam et al., unpublished. See also Harris et al., 1996; McCormack et al., 2020.

68. Bulley & Schacter, 2020; Redshaw & Suddendorf, 2020. In fiction, authors sometimes explore counterfactuals in great detail, such as when entertaining what might have happened if someone had prevented the birth of Adolf Hitler (Fry, 1996).

69. Roberts & Stewart, 2018.

Chapter 4. Under the Hood

1. Tulving, 1985.

2. Neuropsychologists had previously performed extensive brain imaging during Cochrane's life; for a review of his biography and contributions to science, see Rosenbaum et al., 2005. The postmortem analysis confirmed earlier imaging results and identified the extent of Cochrane's extensive medial temporal lobe damage (Gao et al., 2020).

3. For other research that has corroborated the idea that damage to the hippocampus and surrounding structures causes amnesia for the past that also impairs foresight, see Hassabis et al., 2007; Klein et al., 2002; Palombo et al., 2015.

4. Bloom, 2004.

5. Buonomano, 2017.

6. Evidence suggests the visual system gets ahead of the game by representing anticipated future events as if they were already happening, in advance of any sensory input (Blom et al., 2020).

7. Natural selection has produced diverse mechanisms for tracking the passage of time. There is extensive research on how humans track durations and perceive time passing. For instance, see Wittmann, 2016.

8. Nijhawan, 2008.

9. Bar, 2009.

10. Helmholtz, 1866/1925.

11. Barrett & Simmons, 2015; Clark, 2016; Corcoran et al., 2020; Friston, 2010; Rao & Ballard, 1999.

12. Imamoglu et al., 2012; Imamoglu et al., 2013.

13. Seth, 2019.

14. This even occurs at the very elementary levels of the visual system, where features like the orientation of lines get processed, and regardless of whether we are also dealing with competing demands, like a taxing working-memory task (Garrido et al., 2016; Tang et al., 2018).

15. Clark, 2013.

16. Koch, 2016.

17. BBC News, 2019. The story of DeepMind's quest to solve Go is told masterfully in the 2017 documentary *AlphaGo*.

18. Glimcher, 2011; Schultz, 1998; Watabe-Uchida et al., 2017.

19. Some of the earliest attempts to understand how organisms could pursue goals came from the discipline known as cybernetics, a word derived from the Greek word *kubernētēs* meaning "steersman" (Weiner, 1950).

20. Burgess, 2014. For key research that led to the prize, see for instance, Hafting et al., 2005; O'Keefe & Dostrovsky, 1971; O'Keefe & Nadel, 1978.

21. Epstein et al., 2017; Moser et al., 2017.

22. Foster & Wilson, 2007; Johnson & Redish, 2007.

23. Tolman, 1939. This behavior is called *vicarious trial and error* (Muenzinger, 1938).

24. For a thorough review of vicarious trial and error and.associated neural activity, see Redish, 2016.

25. Redish, 2016.

26. Quiroga, 2019.

27. Eichenbaum, 2014; Umbach et al., 2020.

28. Quiroga, 2021.

29. Schacter et al., 2007. An fMRI scanner picks up on the unique magnetic features of blood that is carrying oxygen versus blood that is not. Because brain cells use oxygen as fuel, neuroscientists infer that regions flushed with oxygenated blood are hard at work. For a neuroimaging meta-analysis on the overlap between remembering the past and imagining the future, see Benoit & Schacter, 2015.

30. Addis et al., 2007; Okuda et al., 2003. While the hippocampus is undoubtedly important, research on this region may have overshadowed the role of other brain regions involved in mental time travel. For instance, research shows that the angular gyrus of the brain's parietal lobe appears to be critical for the sensory and perceptual details of imagined scenarios (Ramanan et al., 2018; Thakral et al., 2017). Also note that some regions may be even more active during imagining the future than when remembering the past. This could be because imagining future events requires new connections between mental details rather than just reactivating connections that already exist (Benoit & Schacter, 2015).

31. For details of some of these problems, see Miloyan & McFarlane, 2019; Miloyan, McFarlane, et al., 2019.

32. Aging: Schacter et al., 2018. Dementia: Irish et al., 2013; Irish et al., 2012; Irish & Piolino, 2016.

33. Henry et al., 2016; Lyons et al., 2014; Lyons et al., 2016; Lyons et al., 2019; Terrett et al., 2017.

34. Bulley & Irish, 2018.

35. Self-projection: Buckner & Carroll, 2007; Suddendorf & Corballis, 1997. Mirror self-recognition: Nielsen et al., 2003. Even toddlers can update these expectations quickly. In one of our studies, we placed children in a high chair with a tray to prevent them seeing their legs directly but in front of a mirror so they could see their legs in the reflection. When we surreptitiously placed a sticker on their leg, the children acted in much the same way as they do when marked on their forehead—reaching for the surprising sticker. In another condition, we attached a pair of baggy pants to the chair and slipped the children into them without giving them a chance to see the pants directly. When the sticker was added and the children were presented with the mirror, they ignored the sticker. However, when a final group of children was tested in the same way but given thirty seconds to see the baggy pants directly before we blocked their view by affixing the tray, they passed the task and reached for the sticker. Thirty seconds of exposure was enough for them to update their expectations of what their legs looked like (Nielsen et al., 2006).

36. Gallup, 1970; Suddendorf & Butler, 2013; Butler & Suddendorf, 2014.

37. In one study from our team, led by David Butler in collaboration with Jason Mattingley and others at the Queensland Brain Institute, we found distinct neural signatures depending on whether participants were looking at photos of themselves or at mirror images (Butler et al., 2012). And when we presented twins with photos of either themselves or their twin from different times of their lives, the neural processes associated with recognition were the same—except for the processes typically associated with retrieving memories (Butler et al., 2013).

38. Fleming, 2021. For a review of selfhood in dementia, see Strikwerda-Brown et al., 2019. For other perspectives on selfhood across time, see D'Argembeau, 2020.

39. Canadian Press, 2007.

40. With maturation of mental time travel capacities, children begin to think about death (Slaughter & Griffiths, 2007) and about the possibilities of an afterlife (Bering & Bjorklund, 2004). There has been much speculation about far-reaching consequences of the fear of death (Greenberg et al., 1997).

41. Laukkonen & Slagter, 2021. For a meta-analysis of the brain regions implicated in meditation, see Fox et al., 2016.

42. Buckner & DiNicola, 2019; Smallwood et al., 2021. When blood flows to one of the regions, so too does blood flow to the other regions, suggesting that activity in these different regions tends to co-occur. That's why cognitive neuroscientists call it a network.

43. Galéra et al., 2012; Corballis, 2015; Ruby et al., 2013. For a review of the costs and benefits of mind wandering, see Mooneyham & Schooler, 2013.

44. Seli et al., 2018. For a broader discussion on the distinction between intentional and unintentional mind wandering, see Seli et al., 2016.

45. Burgess et al., 2007; Golchert et al., 2017; Smallwood & Schooler, 2015.

46. O'Callaghan et al., 2019. For a similar finding in healthy aging, see Irish et al., 2019.

47. Burgess et al., 2007; Fleming & Dolan, 2012; Smaers et al., 2017.

48. Gilbert, 2006.

49. Dawes et al., 2020.

50. Hesslow, 2002. For a review of mental imagery, see Pearson, 2019. Evidence suggests that when people are asked to imagine the future, the vast majority of the time (over 90 percent) they do so with mental imagery rather than with inner speech alone (Clark et al., 2020).

51. Miloyan, Bulley, et al., 2019; Sapolsky, 2004.

52. Bulley et al., 2019. For a thorough meta-analysis of the effect of imagining the future on delay discounting, see Rösch et al., in press.

53. The first studies on the effect of imagining the future on delay discounting were fMRI studies (Benoit et al., 2011; Peters & Büchel, 2010), which followed an influential theoretical argument from Boyer (2008).

54. Gilbert & Wilson, 2007.

55. Darwin, 1871.

56. Gilbert et al., 1998.

57. Miloyan & Suddendorf, 2015.

58. Gautam et al., 2017; Kopp et al., 2017.

59. Seneca, 65 AD/1969.

60. LeDoux, 2015; Nesse, 1998. According to some clinicians, such as Adrian Wells, metacognition is among the reasons normal worrying—which may be useful—can become a disorder with debilitating consequences. Some of our biggest problems arise from our worrying about worrying, like when we worry our worrying will make us less fun to be around or drive us mad (Wells, 2005).

61. Craik, 1943. For a review of Craik's seminal but underappreciated contributions to cognitive science, see Williams, 2018.

62. Benoit et al., 2019.

63. Reddan et al., 2018.

64. This was a major theme driving the cognitive revolution in psychology—see Chomsky, 1959.

65. Poldrack, 2018; Dehaene, 2014.

Chapter 5. Are Other Animals Stuck in the Present?

1. The Eden Killer Whale Museum, Australia, tells the full story. See also Crew, 2014.

2. For brain comparisons, see Jerison, 1973; Marino, 2007; Roth & Dicke, 2005. Orcas have one of the heaviest brains, about eleven pounds, but an EQ of "only" 2.5.

3. Marino et al., 2007; Smolker et al., 1997; Visser et al., 2008. Dolphins seem to also play with bubble rings (Janik, 2015).

4. Some birds, such as herons, can also be observed fishing with baits or lures. This is likely learned by associating particular floating objects with the occurrence of fish prey (Ruxton & Hansell, 2011).

5. Kuczaj & Walker, 2006.

6. There are signs of numerical competence in nonhuman animals, but different results tend to emerge if one studies this question by training animals or by examining their spontaneous choices (Agrillo & Bisazza, 2014).

7. Suddendorf, 2013a.

8. Mitchell et al., 2009. See also Tagkopoulos et al., 2008.

9. Lorenz & Tinbergen, 1939.

10. Feeney et al., 2012.

11. Davies et al., 2003.

12. Dawkins et al., 1979. For threat response mechanisms, fear, and anxiety, see Miloyan, Bulley, et al., 2019. Note that any regularity in predation or avoidance behavior also leads to the risk of an opponent evolving the means to exploit this. So the arms race may also select for some measure of unpredictability.

13. Much of evolutionary psychology has focused on the need to attract a sexual partner and the consequences of sexual selection (Brooks, 2011; Buss, 1999; Miller, 2000; Zietsch et al., 2015).

14. Suddendorf et al., 2018.

15. Associative learning is strongest when the co-occurrence of events is surprising (Rescorla & Wagner, 1972; Roesch et al., 2012).

16. Maier & Seligman, 2016. Repeated exposure to stimuli can also be useful for predicting safety. For instance, when two sets of fertile chicken eggs are exposed to different tones, the hatched chicks emit fewer distress calls when again exposed to the sound they had heard while inhabiting the shell (Rajecki, 1974).

17. Ferster & Skinner, 1957; Thorndike, 1898.

18. Brosnan & de Waal, 2003.

19. Engelmann et al., 2017; Ulber et al., 2017; Wynne, 2004.

20. Studies suggest that rats can even adjust what they had learned to expect retrospectively. If a rat feeds on two novel fruits, say an apple and a pear, and then feels happily sated, it may not know which of the two food sources, or both, brought the satiation. If it then feeds on apple and fails to feel sated afterwards, rats will subsequently seek out pear more than control groups do. They have retrospectively increased the value they expect to get from eating pear, even though all they experienced since was an unsatisfying apple (Adams & Dickinson, 1981; Dickinson, 2012).

21. Skinner, 1948.

22. Taste aversion learning has been highlighted as an exception because rats have been documented to learn to avoid tastes that later were followed by sickness (Garcia et al., 1966). However, it may be possible that burps can return a flavor experience that is then directly paired with being unwell.

23. Clayton & Dickinson, 1998; Clayton et al., 2001.

24. Suddendorf & Corballis, 2007.

25. Suddendorf & Busby, 2003.

26. Some researchers maintain a rich interpretation of such results. The psychologist Jonathon Crystal, for instance, argues that animals have "elements of episodic memory." He argues this based on experiments on various aspects of memory in rats. For a debate about the strengths and weaknesses of the evidence, see Crystal & Suddendorf, 2019.

27. Cheng et al., 2016; Suddendorf & Corballis, 2010.

28. For a review, see Bulley et al., 2016.

29. Santos & Rosati, 2015; Tobin & Logue, 1994.

30. Beran & Evans, 2006. These tasks have different demands and may not all measure the same ability (Bulley et al., 2016). In one study, chimpanzees even seemed to distract

themselves with other activities to aid their waiting (Evans & Beran, 2007)—though this may simply reflect their attempt to reduce the agitation of having to wait, rather than any awareness that this strategy would improve their chances of waiting successfully.

31. Goodall, 1986.

32. Janmaat et al., 2016.

33. van Schaik et al., 2013.

34. In early experiments, Edward Tolman established that rats learn the layout of their environment even without reinforcement. Rats that were rewarded when reaching the exit of a maze gradually made fewer wrong turns the more trials they ran. Other rats that were not rewarded did not improve. However, once a reward was introduced, the latter rats moved through the maze just as efficiently as the group that had been rewarded all along (Tolman, 1948).

35. Data on preplay: Dragoi & Tonegawa, 2011; Pfeiffer & Foster, 2013. Debate: Corballis, 2013a, 2013b; Balter, 2013; Bendor & Spiers, 2016; Suddendorf, 2013b.

36. Collias & Collias, 1984.

37. Redshaw & Suddendorf, 2016.

38. Suddendorf et al., 2017; Suddendorf, Watson, et al., 2020.

39. In one study, researchers found that chimpanzees did not consistently use both hands even when the future was certain and a treat was dropped in each of two parallel tubes simultaneously, suggesting that perhaps this is a motor coordination problem after all (Lambert & Osvath, 2018). Still, in about a quarter of the trials, the chimpanzees covered both exits, which is around seven times more than when only one item was randomly dropped into one of the tubes, suggesting the problem is in the mind rather than in the behavioral response. Chimpanzees clearly can cover both exits. In our original study, the chimpanzee Holly even covered two exits on two of the first twelve trials, making us suspect she might figure out the problem. We then gave her dozens of further trials but she only covered both exits two more times. To entice her to use both hands, we deliberately made the food come out of whichever exit she did *not* cover, but after ten such trials she just gave up rather than prepare for both possibilities.

40. A recent study tried to challenge this conclusion (Engelmann et al., 2021). Chimpanzees watched as a researcher first held up a piece of apple and concealed it between both of his hands. The researcher then put each hand inside separate opaque boxes and left the piece of apple in one of them before revealing his empty hands. When the chimpanzees had the opportunity to pull one or both boxes within reach, some tended to pull in both. This suggests that the chimpanzees may be uncertain about something that has already happened, but it does not tell us whether they understand alternative future possibilities. To pass the tube task, participants have to consider two mutually exclusive possibilities before the uncertain event occurs—giving them a chance to prepare accordingly.

41. Studies have shown that gorillas and orangutans can also select appropriately long tools for problems that are out of sight at the time of selection (Mulcahy et al., 2005) and that New Caledonian crows can use a tool to obtain another tool they need (Gruber et al., 2019).

42. Boesch & Boesch, 1984.

43. Osvath, 2009; Osvath & Karvonen, 2012.

44. Bischof, 1985; Bischof-Köhler, 1985; Suddendorf & Corballis, 1997.

45. One study on food-caching jays also challenged this hypothesis (Cheke & Clayton, 2012). The birds were first fed to satiation with one type of food and then given the chance to cache that food in two locations. The jays tended to store the food in the location that would be accessible when they would desire that food again in the future. However, the authors acknowledge that the birds may have associated a motivational preference for the food with the relevant storage location during training. If so, then the storage behavior might have reflected a cued present motivational state, rather than any awareness of a future motivational state (Redshaw, 2014).

46. While this does not test the Bischof-Köhler hypothesis per se (which is about anticipating future drive states like agitation, hunger, or thirst), it is nonetheless an interesting question whether an animal has the foresight to keep a tool they will only be able to use at a later stage.

47. Mulcahy & Call, 2006; Suddendorf, 2006.

48. A second condition involved the raven returning a token for a reward, leading to similar results (Kabadayi & Osvath, 2017).

49. Hampton, 2019; Redshaw et al., 2017. This lean perspective has received powerful empirical support from a study on children (Dickerson et al., 2018).

50. Gruber et al., 2019; Taylor, Elliffe, et al., 2010; Taylor, Medina, et al., 2010; Taylor et al., 2012.

51. Boeckle et al., 2020. This is known as an "adversarial collaboration," a study that involves bringing scholars together who have previously argued for different interpretations. We all agreed on a precise methodology in advance and committed to reporting whatever the findings would be, regardless of how the chips fell.

52. Exchanging tokens: Dufour & Sterck, 2008; Tecwyn et al., 2013. Twisting paddles: Tecwyn et al., 2013. Preparing tools: Bräuer & Call, 2015.

53. Suddendorf, 2013a.

54. For ape working-memory capacity estimates, see Balter, 2010; Read, 2008. Some research has suggested that chimpanzees flexibly update working memory and show susceptibility to distraction, as humans do (Völter et al., 2019). For Ai's memory tasks: Inoue & Matsuzawa, 2007; Kawai & Matsuzawa, 2000; Silberberg & Kearns, 2009.

55. Krupenye et al., 2016. But there remain questions about how to interpret implicit theory of mind tasks and anticipatory-looking data (Kulke & Hinrichs, 2021). See also Povinelli & Vonk, 2003; Tomasello et al., 2003; Penn et al., 2008; Suddendorf, 2013a; Vonk, 2020.

56. The open-endedness in music, language, and number is generally considered the result of recursion, and the apparent absence of recursive capacities in nonhuman animals has been widely discussed (Corballis, 2011; Hauser et al., 2002).

57. Suddendorf & Busby, 2003.

58. Redshaw, 2014; Suddendorf, 1999.

59. Conceiving of representations as representations is known as a metarepresentation (Perner, 1991).

60. Redshaw & Bulley, 2018.

61. Delsuc, 2003.

62. Huxley in the second edition of Darwin (1871).

63. Seligman et al., 2016.

64. Goodall, 1986.

65. With Michael Corballis, Thomas proposed that there is something distinct about human mental time travel (Suddendorf & Corballis, 1997), and he and examined various possibilities—including the hypothesis that nonhuman animals' foresight is restricted by deficits in any one, or any combination, of the components illustrated with the theater metaphor (Suddendorf & Corballis, 2007).

66. Cross & Jackson, 2019.

67. Godfrey-Smith, 2016.

Chapter 6. Discovery of the Fourth Dimension

1. Goren-Inbar et al., 2004. See also Shahack-Gross et al. (2014) for evidence of later repeated use of a central hearth, around three hundred thousand years ago.

2. Annaud, 1981.

3. It is difficult detective work to distinguish remnants of controlled fires from wildfires in sediments, let alone to distinguish control of fire from making fire. Some excavations, such as at Wonderwerk Cave in South Africa, suggest potential fire use even earlier than Gesher Benot Ya'aqov. But evidence of widespread use in Africa, Asia, and Europe, and so more likely reflecting the actual making of fires, emerges "only" from about three hundred thousand years ago. For evidence from the Wonderwerk cave, see Berna et al., 2012, and for an overview of current knowledge about the prehistory of fire use, see Gowlett, 2016.

4. Wrangham, 2009.

5. It used to be thought that birch bark production was a very complicated process, but more recent work suggests otherwise (Kozowyk et al., 2017; Schmidt et al., 2019).

6. Taungurung Land and Waters Council, 2021.

7. Suddendorf, 1994.

8. The earliest anatomically modern *Homo sapiens* fossils used to be two hundred thousand years old (Leakey, 1969; McDougall et al., 2005), but some evidence suggest that our species may go back even as far as another hundred thousand years or so (Hublin et al., 2017).

9. Brown et al., 1985. For comparisons of cranial capacities, see Holloway, 2008; Robson & Wood, 2008.

10. The *Homo erectus* migration has long been estimated to have occurred about 1.8 million years ago, with early fossils from Java, China, and Georgia. A more recent report, however, suggests an even earlier date of 2.1 million years for stone tools from China (Zhu et al., 2018).

11. This is sometimes called the *East side story* (Coppens, 1994).

12. Velvet worm squirting (Read & Hughes, 1987).

13. Suddendorf, 2013a. It has been argued that the emergence of aimed throwing played a role in the evolution of the human brain (Calvin, 1982) and cooperation (Bingham, 1999).

14. Young, 2003.

15. Harmand et al., 2015; Lewis & Harmand, 2016.

16. Tool-making replication studies suggest that individuals copied the method of Oldowan tool production from each other (Stout et al., 2019).

17. For a summary of Oldowan tools and their cognitive implications, see Toth & Schick, 2018. But also see Osvath & Gardenfors, 2005.

18. Pargeter et al., 2019.

19. The possibility that these tools could have been made without much planning has been raised (Corbey et al., 2016; Rogers et al., 2016), but it is difficult to see how reinforcement learning or innate fixed action patterns could result in the manufacture and use of such complex technology. Some social learning, at least through imitation, appears to be entailed (Shipton, 2020).

20. Hallos, 2005.

21. Water flow, downslope movement, and other natural processes can affect artifact densities observed today (Potts et al., 1999), so we cannot entirely rule out that the dense accumulation of tools is the product of some kind of natural deposition.

22. Suddendorf et al., 2016.

23. Bramble & Lieberman, 2004. Note, though, that not every find of *Homo erectus* is associated with Acheulean tools. The tools first appear in Africa, and early *Homo erectus* of Java and China appear not to have had these tools; instead, the technology arrived later, with the earliest Asian finds currently known from India (Dennell, 2011).

24. Brooks et al., 2018. The research also found evidence these hominins carried ocher pigment with them from dozens of miles away. Ocher is often associated with cave paintings, body painting, and other artistic endeavors.

25. There is some controversy about the precise definition of the *Homo heidelbergensis* taxon, and there are proposals to recategorize the fossils (Roksandic et al., 2022).

26. White & Ashton, 2003.

27. There is compelling evidence of the use of composite tools from around three hundred thousand years ago (e.g., Sahle et al., 2013), but this technology may have started considerably earlier (Wilkins et al., 2012).

28. Ambrose, 2010.

29. Thieme, 1997.

30. Milks et al., 2019.

31. Conard et al., 2020.

32. Ortega Martínez et al., 2016.

33. Eventually, humans clearly prepared in advance as they dug pits to trap even the biggest giants. In 2019, two large traps in Mexico were found containing the remains of at least a dozen slaughtered woolly mammoths. But those are from fifteen thousand years ago, rather than at the transition from the Acheulean to the Levallois technology hundreds of thousands of years earlier (Tuckerman, 2019).

34. Evidence from Qesem cave in Israel suggests that around this time, stone tool cores were shared, such that skilled knappers appear to have deliberately provided learning opportunities to others less experienced than themselves. The evidence comes from an assembly of hundreds of cores at Qesem cave. Many cores show considerable skill and sophistication at the first stage of production and then, at the second stage, show surprising rookie mistakes. This has been interpreted as reflecting skilled knappers sharing the prepared cores with novices to give them the opportunity to learn (Assaf, 2019).

35. Ambrose, 2010.

36. Suddendorf, Kirkland, et al., 2020.

37. Falk, 2009.

38. Langley & Suddendorf, 2020.

39. Hardy et al., 2020.

40. There is evidence that some humans migrated earlier to just outside Africa, perhaps as early as two hundred thousand years ago, including to Israel (Hershkovitz et al., 2018) or even to Greece (Harvati et al., 2019).

41. van den Bergh et al., 2016.

42. Ikeya, 2015. Supporting the case for seafaring is an analysis of ancient obsidian artifacts, only a few thousand years later, from Sinbuk in Korea, which has shown that they originated in Japan (Lee & Kim, 2015).

43. Unsurprisingly, remains of watercraft are not likely to survive tens of thousands of years. The oldest current hard evidence is the Mesolithic "Pesse canoe," found in the Netherlands and dated to about ten thousand years old (Wierenga, 2001).

44. McBrearty & Brooks, 2000.

45. Corballis, 2017.

46. For instance, see Nobel & Davidson, 1996.

47. Flutes are over thirty-five thousand years old (Conard et al., 2009). The prime candidate for the oldest visual map is currently a twenty-seven-thousand-year-old engraved mammoth tusk that appears to depict rivers, valleys, and routes around ancient Pavlov in what is now the Czech Republic. For a review of maps in prehistoric art, see Utrilla et al., 2021. Sometimes maps also include instructions about what to do where. The cover image of Thomas's previous book, *The Gap*, includes part of a painting from the Cova dels Cavalls in Spain, which is a rock shelter on the bank of Valltorta ravine. It depicts how hunters drive deer into a dead end, where other hunters wait in position ready to kill the trapped animals. It was selected by the publisher for aesthetic reasons, and so Thomas was pleasantly surprised when, upon visiting the site, he discovered the apparent instruction on how to hunt in the valley below.

48. Davies, 2021.

49. Clarkson et al., 2017.

50. Yellen et al., 1995.

51. Grun et al., 2005.

52. Blombos cave finds: Henshilwood et al., 2004; Henshilwood et al., 2009; Henshilwood et al., 2018. Suggestion that cross-hatched patterns represent weaving instructions: Anderson, 2020.

53. Bar-Yosef Mayer et al., 2020.

54. Aubert, Setiawan, et al., 2018.

55. Langley et al., 2020.

56. Burials: Pomeroy et al., 2020. Necklace: Finlayson et al., 2019; Radovčić et al., 2015. Cave art: archaeologists in Spain have discovered handprints and a series of painted shapes that may be sixty-five thousand years old. If so, this would be more than twenty thousand years before modern humans made it to Europe and considerably older than any other known cave art from anywhere in the world (Hoffmann et al., 2018), but for a skeptical take on this evidence, see White et al., 2020, and Aubert, Brumm, et al., 2018. Another find associated with Neanderthals in Germany over fifty thousand years ago included engraved bones (Leder et al., 2021).

57. Wallace, 1870.

58. Browne, 2013.

59. Bonta et al., 2017.

60. Wiessner, 2014.

61. Donald, 1999.

62. Corballis, 2002.

63. Tomasello, 2019.

64. Goren-Inbar et al., 1994; von Hippel, 2018.

65. Shipton & Nielsen, 2015.

66. Lordkipanidze et al., 2005.

67. Bonmatí et al., 2010.

68. Boyer, 2008. For a review of the relationship between theory of mind and mental time travel, see Gaesser, 2020.

69. Price, 2019.

70. Barnosky et al., 2011.

71. De Vos et al., 2015; Di Marco et al., 2018; Kolbert, 2014.

72. While the arrival of humans in New Zealand, Mauritius, and Cyprus is clearly associated with the extinction of local fauna, in other cases the link is less certain (Louys et al., 2021).

73. Flannery, 1994.

74. McWethy et al., 2009.

75. Détroit et al., 2019.

76. Hobbes, 1651; Pinker, 2011.

77. Goodall, 1986. See discussion in Suddendorf, 2013a. Recently it has been reported, for the first time, that chimpanzees also cooperated in killing another species of ape, a gorilla (Southern et al., 2021). For more on warfare from an evolutionary perspective, see Majolo, 2019.

78. von Hippel, 2018.

79. Bingham, 1999.

80. Frank, 1988.

81. Buckner & Glowacki, 2019. For example, analyses of violence in forty-four traditional South American societies found that in only 2 percent (5 out of 238) of death events did an attacker die (Walker & Bailey, 2013).

82. Sun Tzu, ca. fifth century BC/1910.

83. Sala et al., 2015.

84. Keeley, 1996; Lahr et al., 2016; Thorpe, 2003.

85. Hershkovitz et al., 2018; Stringer et al., 1989.

86. Green, Krause, et al., 2010.

87. Denisovan admixture: Reich et al., 2011. New evidence suggests a group in the Philippines has an even higher level of Denisovan ancestry (Larena et al., 2021).

88. Suddendorf, 2013a.

Chapter 7. Travel Tools

1. Bertemes & Northe, 2007.

2. Mukerjee, 2003. The trajectory reverses when the Earth's pole is tilted farthest away from or towards the sun.

3. Wurdi Youang: Norris et al., 2013. Nabta Playa: Malville et al., 2007; Malville et al., 1998. Stonehenge: Ruggles, 1997. Signs of possible ancient astronomical structures have also been discovered in Peru (Ghezzi & Ruggles, 2007) and many other locations.

4. Gaffney et al., 2013. A case has been made for even earlier lunar calendars (Marshack, 1972).

5. Orwig, 2015. There is some debate about the precise age of the disk (Gebhard & Krause, 2020; Pernicka et al., 2020).

6. Copernicus, 1543/1976. Similar models of the solar system had been earlier proposed in ancient Greece and the Middle East, and it is debated whether Copernicus was aware of these works (Swerdlow & Neugebauer, 1984).

7. Sánchez et al., 1999.

8. Uetz et al., 1994.

9. Biological clocks drive bird migration (Gwinner, 2003), chipmunk hibernation (Kondo et al., 2006), and cicada swarming (Gould, 1992).

10. Clark & Chalmers, 1998; Sutton, 2010.

11. Yolngu: Clarke, 2009; Green, Billy, et al., 2010. Inuit: Poirier, 2007.

12. Bradley & Yanyuwa families, 2010.

13. Chatwin, 1987; Kelly, 2016. One of the most ancient songlines traces the journey of the seven sisters that make up the star cluster Pleiades. The songline covers more than half the width of Australia, from the central desert to the west coast, providing vital information about water holes and other critical features of the land. The sisters were pursued by a lecherous man and ended up launching themselves from a steep hill into the sky. But the man followed them and became one of the stars in Orion's belt. This corresponds to the rise of Pleiades in the night sky followed by the constellation of Orion.

14. This may have been more widespread than often thought. In his book *Dark Emu*, Bruce Pascoe (2014) draws on records and diaries of early explorers to make the case that Aboriginal Australians had long been domesticating plants, storing harvests, and using firestick farming to shape the landscape. Pascoe's claims have been questioned by anthropologist Peter Sutton and archaeologist Keryn Walshe (2021). They contend that, although some early Aboriginal Australians were aware of the possibility of agriculture, these people almost universally rejected farming in favor of hunter-gatherer lifestyles.

15. Scott, 2017. For a sweeping account of the rise of agriculture since the last ice age and the role of geographic factors in its spread, see Diamond, 1997.

16. Ritchie, 2019.

17. Bar-On et al., 2018.

18. Mukerjee, 2003. The Nebra sky disk may also have had agricultural uses. The Pleiades constellation routinely appears at dawn in northern skies in the fall season, before disappearing at dawn in spring—marking the ideal times for harvesting and planting.

19. Britton, 1989; Lehoux, 2007; Steele, 2017.

20. Steele, 2021.

21. Hunger & Steele, 2018.

22. Saturno et al., 2006.

23. Vail, 2006.

24. Bricker et al., 2001; Teeple, 1926.

25. Thompson's (1972) translation.

26. Duncan, 2011.

27. Suetonius, 121 AD/1957.

28. Macrobius, 431 AD/2011.

29. Berndt et al., 1993.

30. Pliny the Elder (77 AD/1855) comments, "An obelisk is a symbolic representation of the sun's rays, and this is the meaning of the Egyptian word for it" (Section 36.14).

31. Al-Rawi & George, 1991.

32. Hunger & Steele, 2018.

33. al-Jazari, 1206/1974.

34. Addomine et al., 2018.

35. Willms et al., 2017.

36. Norris & Owens, 2015.

37. Samuelson, 2019.

38. The twenty-four-hour day appears to have had its origins in Babylon, where people separated their sundials into twelve sections and added another twelve hours to represent the night (Dohrn-van Rossum, 1992/1996). The base-60 numeral system was developed by the Sumerians and popularized by the Babylonians, and in 1000 AD, Iranian scholar al-Bīrūnī adopted this system to define the minute and second (al-Bīrūnī, 1000/1879).

39. Donnachie, 2000. Other numbers cannot be so straightforwardly subdivided. The decimal system, for instance, is in some ways less practical because 100 can be wholly divided into 2×50, 4×25, and 5×20, but not by 3 or 6.

40. Igbo: Thomas, 1924. Javanese: Ammarell, 1996. Akan: Bartle, 1978. Romans: Ker, 2010. Chinese: Smith, 2011. Egyptians: Parker, 1974. French: Shaw, 2011. The seven-day revolving week: Zerubavel, 1989.

41. Cicero, 44 BC/1853.

42. Cicero, 44 BC/1923.

43. Hunger & Steele, 2018. Another, more ancient Babylonian text, the *Enuma Anu Enlil*, consists of around seven thousand similar celestial omens spread across seventy tablets (Chadwick, 1984).

44. The period of 260 days fails to directly map onto any celestial cycles, but it does neatly fit three times into the Mayans' measured Mars cycle (780 days).

45. Thompson's (1972) translation.

46. Writing may have also been independently invented in the Easter Islands. The *rongorongo* script found there is yet to be deciphered, although one section appears to contain a type of lunar calendar (Horley, 2011).

47. Chou, 1979.

48. Cicero, 44 BC/1923.

49. Reagan: Weisberg, 2018. Houdini: Puglise, 2016.

50. Basu et al., 2009.

51. Another simple way to keep track is a "tally stick" such as a bone or log with notches carved into it. Several apparent tally sticks have been discovered in the archaeological record, including the Lebombo bone from southern Africa, which has been dated to over forty thousand years old (d'Errico et al., 2012). Running the length of this baboon fibula are twenty-nine notches: it has been speculated that someone was using it to keep track of the

lunar cycle, or a menstrual cycle (Darling, 2004). However, it would be an understatement to point out that the function of these objects is difficult to determine.

52. Boltz, 2000.

53. Laozi (traditional), ca. fourth to sixth century BC.

54. Unknown author, ca. ninth century BC, as translated by Jacobsen (1983).

55. Tyerman & Bennet, 1831.

56. Best, 1921.

57. Blust, 1996; Matisoo-Smith, 2015.

58. Jacobsen, 1983.

59. Ascher & Ascher, 1997.

60. Clindaniel, 2019.

61. Urton, 2003.

62. Unknown author, thirty-first century BC. Note that the Schøyen Collection gives the modern metric of 134,813 liters rather than a direct translation of the number (29,086), and also gives an alternative translation of the name (Kushin). It remains possible that Kushim is a more generic title for the holder of a particular office, and that the tablet is not an order but a receipt. Indeed, an alternative interpretation noted by Harari (2014) is: "A total of 29,086 measures of barley were received over the course of 37 months. Signed, Kushim." In any case, signed records were kept for the future.

63. Harari, 2014.

64. The opening word *If* is used in modern translations to show how these laws would have been originally interpreted (King, 1910).

65. Bryce, 2006.

66. Aristotle, ca. 328 BC/2008. Chapters 63–69 describe various randomization and anonymization procedures of the Athenian court.

67. Boyer, 2020.

68. Aristotle, ca. 328 BC/2008.

69. Borges, 1967.

70. Berg, 1992; Fitzpatrick & McKeon, 2020.

71. Rai stones: Bruce et al., 2000. Pay the bearer: Bordo & Schwartz, 1984. Money as trust: Ferguson, 2008.

72. Plato (ca. 360 BC), as quoted in Hackford's (1952) translation, 274c–275b. Note that Plato portrayed Socrates as quoting (and endorsing) the words supposedly spoken by the Egyptian king, Thamus.

73. Carr, 2008.

74. Ruginski et al., 2019; Lu et al., 2020.

Chapter 8. Our Slice of Time

1. BBC News, 2020. Only two of the zoo's chimpanzees, Bally and Limbo, managed to survive the inferno and were rescued by firefighters. The mother and her daughters from Krefeld who released the sky lanterns felt deep remorse over the conflagration at the ape house and turned themselves in to the authorities.

2. Herschel, 1830.

3. Newton, 1687.

4. Bristow, 2017.

5. Al-Haytham's most well-known work is the *Book of Optics* (1011/1989), but his passage about the role of skepticism in scientific thinking comes from the opening paragraph of his *Dubitationes in Ptolemaeum* (as noted in Sabra's 1989 translation of the *Book of Optics*).

6. Gribbin & Gribbin, 2017. For an entertaining journey through the human side of some of the major scientific breakthroughs, see Bryson, 2004.

7. Seneca (64 AD/2017) wrote about the relationship between the moon and the tides in *On Providence*: "In point of fact, their growth is strictly allotted; at the appropriate day and hour they approach in greater volume or less according as they are attracted by the lunar orb, at whose sway the ocean wells up."

8. Newton, 1687.

9. Parker, 2011.

10. ATLAS Collaboration et al., 2012.

11. Cicero, 44 BC/1853; Laplace, 1814/1951.

12. Newton, 1687.

13. Einstein penned this phrase while defending determinism in a 1926 letter to fellow physicist Max Born.

14. For a beginner's introduction to the various interpretations of quantum physics, see Bell, 1992. Note that all interpretations require bold assumptions. The pilot wave theory, for instance, requires the assumption of nonlocality, such that particles must influence each other faster than the speed of light.

15. Skinner, 1953.

16. Metaphysics aside, the sense of free will may depend fundamentally on our ability to foresee multiple possible futures (Alexander, 1989).

17. Goldstone, 2000.

18. Bar-On et al., 2018. Even compared with the whole phylum of cordata—all animals with a backbone—arthropods have a greater biomass.

19. Vince, 2019; Zalasiewicz et al., 2017.

20. Behbehani, 1983; Johnson, 2021.

21. Dunsworth & Buchanan, 2017.

22. Harre, 2018. Some researchers, such as the physician Hans Rosling and the psychologist Steven Pinker, have argued that the world has gradually become a significantly better place for a great swathe of humanity, with negatives such as hunger, child mortality, and poverty declining and positives such as sanitation, immunizations, and literacy rising (Pinker, 2018; Rosling et al., 2018).

23. The earthquake was so forceful that it sped up the Earth's rotation and shortened our days by a few millionths of a second (Hopkin, 2004).

24. Hettiarachchi, 2018.

25. For the misprediction of Velpeau: Velpeau, 1839; Eger II et al., 2014. For Thomson, Einstein, and Lewyt: Navasky, 1996. And for Summerfield: Albrecht, 2014.

26. McNeill, 2001.

27. Alas, the predicted growth in plastic waste still exceeds efforts to mitigate plastic pollution (Borrelle et al., 2020).

28. Laville & Taylor, 2017.

29. Napper et al., 2020; Peng et al., 2018.

30. Law & Thompson, 2014.

31. Estrada et al., 2017.

32. Suddendorf, 2013a. But groups such as the UNEP Great Ape Survival Partnership and the Jane Goodall Institute are working hard to change the prospects and save our closest animal relatives.

33. One scientific report concluded: "Dwindling population sizes and range shrinkages amount to a massive anthropogenic erosion of biodiversity and of the ecosystem services essential to civilization" (Ceballos et al., 2017).

34. Kolbert, 2014.

35. See Kolbert (2014) for the story of Cuvier's discovery of extinction. Of course, only with the knowledge of extinction can we plan to cause extinctions. We may, for instance, soon be able to completely eradicate entire species of mosquito using genetic engineering and other high-tech methods. There are reasons why people may want to pursue such a radical strategy. These bloodsuckers are responsible for a gluttony of human suffering, bringing malaria and other terrible afflictions to countless people. Then again, who knows what vital, as-yet-unappreciated role mosquitoes might play in the fragile machinations of the ecosystem?

36. Even some of the most common fish have been decimated, as engagingly illustrated in Mark Kurlansky's book *Cod: A Biography of the Fish That Changed the World* (1997).

37. This was the 2011–2020 target. This target was not reached by 2020, even if many signatories have been taking steps in the right direction (Secretariat of the Convention on Biological Diversity, 2020).

38. Estimates from Australian Government Great Barrier Reef Marine Park Authority.

39. Ball et al., 2009; Possingham, 2016; Possingham et al., 2000.

40. Dudgeon et al., 2018; Noad et al., 2020; Noad et al., 2019.

41. Ritchie & Roser, 2018.

42. Steffen et al., 2015.

43. Sharot, 2011.

44. Sharot & Garrett, 2016.

45. Szpunar & Schacter, 2013. See also Baumeister et al., 2018.

46. Kahneman, 2011; Tversky & Kahneman, 1973. Daniel Kahneman received a Nobel Prize in Economics in 2002 for his work with Amos Tversky, who by that point had died and so, given that the prize is not awarded posthumously, he could not receive this recognition. For Hofstadter's Law, see Hofstadter, 1979.

47. For base rates and frequencies, see Gigerenzer, 1998; for superforecasters, see Tetlock & Gardner, 2016.

48. Halley, 1693.

49. Hershfield et al., 2018.

50. Lempert et al., 2019.

51. Hume, 1739.

52. Soutschek et al., 2017.

53. Klein, 2007.

54. Elster, 2000.

55. Jones, 1985.

56. Nostradamus (1555) published a collection of four-liners that have since been interpreted (retrospectively) as foretelling wars, floods, earthquakes, the rise of Hitler, and the

bombing of Hiroshima—Nostradamus may have borrowed ideas from classic works and even simply opened books at random to find his inspiration.

57. For example, Toby Ord, extrapolating from earlier natural disasters, puts the chances of an asteroid or comet causing such a tragedy at one in a million, while the existential risk from supervolcanic eruption sits at one in ten thousand. The estimated risks from nuclear war or climate change are one in a thousand apiece. Note that the definition of existential risk here is not restricted to the extinction of the human species but also includes the more nebulous concept of the destruction of humanity's "longer-term potential" (Ord, 2020).

58. Bostrom, 2002.

59. Witze, 2021.

60. Dennett, 1999.

61. Sagan, 1994; if you're looking for something more to kick start your long-term imagination, consider that a party organized by Michael Ogden and Peter Dean (who also designed the invitations to Stephen Hawking's party for time travelers from the future) is set to start at noon on June 6, 2269 (McRobbie, 2021).

62. Some thinkers see the welfare of billions upon billions of prospective people as thumbing the scales of justice away from the welfare of those of us perched in the present. But, then again, maybe precisely because these future generations are so uncertain, we should discount the importance of outcomes the further away they are in time, meaning the lives of people further away in time will be worth less and less relative to tangible outcomes in the here and now (Parfit, 1984).

63. Krznaric, 2020.

64. Some of the planning is threatening to fall tragically short. Just consider the story of the deteriorating concrete dome that was placed over nuclear waste in Enewetak atoll of the Marshall Islands (Rust, 2019).

65. Nuclear Energy Agency, 2019.

66. Spencer, 1862.

67. Optimism: Venkataraman, 2019. Sharing optimistic predictions: Solda et al., 2020; Suddendorf, 2011. Placebo effect: Humphrey, 2000.

68. Suddendorf, 2013a.

69. More, 1516; Skinner, 1969.

70. As Elizabeth Kolbert put it in her book *The Sixth Extinction*: "A hushed hundred million years from now, all that we consider to be the great works of man—the sculptures and the libraries, the monuments and the museums, the cities and the factories—will all be compressed into a layer of sediment not much thicker than a cigarette paper?" (Kolbert, 2014). Just as in Wells's *The Time Machine*, where the time traveler encounters strange human descendant species, the Eloi and Morlock, at some point future generations are likely to be just as different from us as we are from our ancient hominin ancestors (Wells, 1895).

71. Mack, 2020.

REFERENCES

Adams, C. D., & Dickinson, A. (1981). Instrumental responding following reinforcer devaluation. *Quarterly Journal of Experimental Psychology Section B, 33*, 109–121.

Addis, D. R., Wong, A. T., & Schacter, D. L. (2007). Remembering the past and imagining the future: Common and distinct neural substrates during event construction and elaboration. *Neuropsychologia, 45*, 1363–1377.

Addis, D. R., Wong, A. T., & Schacter, D. L. (2008). Age-related changes in the episodic simulation of future events. *Psychological Science, 19*, 33–41.

Addomine, M., Figliolini, G., & Pennestrì, E. (2018). A landmark in the history of non-circular gears design: The mechanical masterpiece of Dondi's astrarium. *Mechanism and Machine Theory, 122*, 219–232.

Aeschylus. (1926). *Prometheus bound* (H. Weir Smyth, Trans.). Harvard University Press. (Original work published ca. 525–456 BC)

Agrillo, C., & Bisazza, A. (2014). Spontaneous versus trained numerical abilities. A comparison between the two main tools to study numerical competence in non-human animals. *Journal of Neuroscience Methods, 234*, 82–91.

al-Bīrūnī, A. R. (1879). *The chronology of ancient nations* (E. Sachau, Trans.). W. H. Allen & Co. (Original work published 1000)

Albrecht, K. (2014). Deconstructing the future: Seeing beyond "magic wand" predictions. *The Futurist, 48*, 44.

Alexander, R. D. (1989). Evolution of the human psyche. In P. Mellars & C. Stringer (Eds.), *The human revolution: Behavioral and biological perspectives on the origins of modern humans*. Princeton University Press.

al-Haytham, I. (1989). *The optics of al-Haytham* (A. I. Sabra, Trans.). W. S. Maney and Son Limited. (Original work published 1011)

al-Jazari, I. a.-R. (1974). *The book of knowledge of ingenious mechanical devices* (D. R. Hill, Trans.). D. Reidel Publishing Co. (Original work published 1206)

Alloway, T. P., Gathercole, S. E., & Pickering, S. J. (2006). Verbal and visuospatial short-term and working memory in children: Are they separable? *Child Development, 77*, 1698–1716.

Al-Rawi, F. N., & George, A. R. (1991). Enūma Anu Enlil XIV and other early astronomical tables. *Archiv für Orientforschung*, 52–73.

Ambrose, S. H. (2010). Coevolution of composite-tool technology, constructive memory, and language: Implications for the evolution of modern human behavior. *Current Anthropology, 51*, S135–S147.

Ammarell, G. (1996). The planetarium and the plough: Interpreting star calendars of rural Java. *Archaeoastronomy, 12*, 320–335.

Anderson, H. (2020). Impressions and expressions: Searching for the origins of basketry. In T. A. Heslop & H. Anderson (Eds.), *Basketry and beyond: Constructing culture*. Brill.

Annaud, J.-J. (Director). (1981). *Quest for fire* [Film]. 20th Century Fox.

Aristotle. (2008). *The Athenian constitution* (F. G. Kenyon, Trans.). Project Gutenberg. (Original work published ca. 328 BC)

Armitage, K. L., Bulley, A., & Redshaw, J. (2020). Developmental origins of cognitive offloading. *Proceedings of the Royal Society B, 287*, 20192927.

Armitage, K. L., Taylor, A. H., Suddendorf, T., & Redshaw, J. (2022). Young children spontaneously devise an optimal external solution to a cognitive problem. *Developmental Science, 25*, e13204.

Ascher, M., & Ascher, R. (1997). *Mathematics of the Incas: Code of the Quipu*. Dover Publications.

Assaf, E. (2019). Core sharing: The transmission of knowledge of stone tool knapping in the Lower Palaeolithic, Qesem Cave (Israel). *Hunter Gatherer Research, 3*, 367–399.

Astell, M. (1697). *A serious proposal to the ladies, for the advancement of their true and greatest interest: In two parts*. Richard Wilkin.

Atance, C. M. (2015). Young children's thinking about the future. *Child Development Perspectives, 9*, 178–182.

Atance, C. M., & Meltzoff, A. N. (2005). My future self: Young children's ability to anticipate and explain future states. *Cognitive Development, 20*, 341–361.

ATLAS Collaboration, Aad, G., Abajyan, T., Abbott, B., Abdallah, J., Khalek, S. A., . . . Zwalinski, L. (2012). Observation of a new particle in the search for the Standard Model Higgs boson with the ATLAS detector at the LHC. *Physics Letters B, 716*, 1–29.

Attenborough, D. (2020). *A life on our planet: My witness statement and a vision for the future*. Grand Central Publishing.

Aubert, M., Brumm, A., & Huntley, J. (2018). Early dates for "Neanderthal cave art" may be wrong. *Journal of Human Evolution, 125*, 215–217.

Aubert, M., Setiawan, P., Oktaviana, A. A., Brumm, A., Sulistyarto, P. H., Saptomo, E. W., . . . Brand, H. E. A. (2018). Palaeolithic cave art in Borneo. *Nature, 564*, 254–257.

Augenblick, N., Cunha, J. M., Dal Bó, E., & Rao, J. M. (2016). The economics of faith: Using an apocalyptic prophecy to elicit religious beliefs in the field. *Journal of Public Economics, 141*, 38–49.

Baddeley, A. (1992). Working memory. *Science, 255*, 556–559.

Ball, I. R., Possingham, H. P., & Watts, M. (2009). Marxan and relatives: software for spatial conservation prioritisation. *Spatial Conservation Prioritisation: Quantitative Methods and Computational Tools, 14*, 185–196.

Balter, M. (2010). Did working memory spark creative culture? *Science, 328*, 160–163.

Balter, M. (2013). Can animals envision the future? Scientists spar over new data. *Science, 340*, 909.

Balter, M. (2015, January 13). Human language may have evolved to help our ancestors make tools. *Science News.*

Bar, M. (2009). The proactive brain: Memory for predictions. *Philosophical Transactions of the Royal Society B, 364*, 1235–1243.

Barnosky, A. D., Matzke, N., Tomiya, S., Wogan, G. O., Swartz, B., Quental, T. B., . . . Maguire, K. C. (2011). Has the Earth's sixth mass extinction already arrived? *Nature, 471*, 51–57.

Bar-On, Y. M., Phillips, R., & Milo, R. (2018). The biomass distribution on Earth. *Proceedings of the National Academy of Sciences, 115*, 6506–6511.

Baron-Cohen, S., O'Riordan, M., Stone, V., Jones, R., & Plaisted, K. (1999). Recognition of faux pas by normally developing children and children with Asperger syndrome or high-functioning autism. *Journal of Autism and Developmental Disorders, 29*, 407–418.

Barrett, L. F., & Simmons, W. K. (2015). Interoceptive predictions in the brain. *Nature Reviews Neuroscience, 16*, 419–429.

Bartle, P. F. (1978). Forty days: The Akan calendar. *Africa, 48*, 80–84.

Bartlett, F. C. (1932). *Remembering: A study in experimental and social psychology.* Cambridge University Press.

Bar-Yosef Mayer, D. E., Groman-Yaroslavski, I., Bar-Yosef, O., Hershkovitz, I., Kampen-Hasday, A., Vandermeersch, B., . . . Weinstein-Evron, M. (2020). On holes and strings: Earliest displays of human adornment in the Middle Palaeolithic. *PLOS ONE, 15*, e0234924.

Basu, S., Dickhaut, J., Hecht, G., Towry, K., & Waymire, G. (2009). Recordkeeping alters economic history by promoting reciprocity. *Proceedings of the National Academy of Sciences, 106*, 1009–1014.

Bauer, P. (2007). *Remembering the times of our lives: Memory in infancy and beyond.* Laurence Erlbaum Associates.

Baumeister, R. F., Maranges, H. M., & Sjåstad, H. (2018). Consciousness of the future as a matrix of maybe: Pragmatic prospection and the simulation of alternative possibilities. *Psychology of Consciousness: Theory, Research, and Practice, 5*, 223–238.

BBC News. (2019, November 27). Go master quits because AI "cannot be defeated." *BBC News.*

BBC News. (2020, January 2). Krefeld zoo fire: German police suspect three women. *BBC News.*

Beaty, R. E., Seli, P., & Schacter, D. L. (2019). Network neuroscience of creative cognition: Mapping cognitive mechanisms and individual differences in the creative brain. *Current Opinion in Behavioral Sciences, 27*, 22–30.

Beck, S. R., Apperly, I. A., Chappell, J., Guthrie, C., & Cutting, N. (2011). Making tools isn't child's play. *Cognition, 119*, 301–306.

Beck, S. R., Robinson, E. J., Carroll, D. J., & Apperly, I. A. (2006). Children's thinking about counterfactuals and future hypotheticals as possibilities. *Child Development, 77*, 413–426.

Behbehani, A. M. (1983). The smallpox story: Life and death of an old disease. *Microbiological Reviews, 47*, 455–509.

Bélanger, M. J., Atance, C. M., Varghese, A. L., Nguyen, V., & Vendetti, C. (2014). What will I like best when I'm all grown up? Preschoolers' understanding of future preferences. *Child Development, 85*, 2419–2431.

Bell, J. (1992). Six possible worlds of quantum mechanics. *Foundations of Physics, 22*, 1201–1215.

Bendor, D., & Spiers, H. J. (2016). Does the hippocampus map out the future? *Trends in Cognitive Sciences, 20*, 167–169.

Benoit, R. G., Gilbert, S. J., & Burgess, P. W. (2011). A neural mechanism mediating the impact of episodic prospection on farsighted decisions. *Journal of Neuroscience, 31*, 6771–6779.

Benoit, R. G., Paulus, P. C., & Schacter, D. L. (2019). Forming attitudes via neural activity supporting affective episodic simulations. *Nature Communications, 10*, 2215.

Benoit, R. G., & Schacter, D. L. (2015). Specifying the core network supporting episodic simulation and episodic memory by activation likelihood estimation. *Neuropsychologia, 75*, 450–457.

Beran, M. J., & Evans, T. A. (2006). Maintenance of delay of gratification by four chimpanzees (*Pan troglodytes*): The effects of delayed reward visibility, experimenter presence, and extended delay intervals. *Behavioural Processes, 73*, 315–324.

Berg, M. L. (1992). Yapese politics, Yapese money and the Sawei tribute network before World War I. *Journal of Pacific History, 27*, 150–164.

Bering, J. M., & Bjorklund, D. F. (2004). The natural emergence of reasoning about the afterlife as a developmental regularity. *Developmental Psychology, 40*, 217.

Berna, F., Goldberg, P., Horwitz, L. K., Brink, J., Holt, S., Bamford, M., & Chazan, M. (2012). Microstratigraphic evidence of in situ fire in the Acheulean strata of Wonderwerk Cave, Northern Cape province, South Africa. *Proceedings of the National Academy of Sciences, 109*, E1215–E1220.

Bernard, E., & Titov, V. (2015). Evolution of tsunami warning systems and products. *Philosophical Transactions of the Royal Society A: Mathematical, Physical and Engineering Sciences, 373*, 20140371.

Berndt, R. M., Berndt, C. H., & Stanton, J. E. (1993). *A world that was: The Yaraldi of the Murray River and the Lakes, South Australia*. UBC Press.

Bertemes, F., & Northe, A. (2007). Der Kreisgraben von Goseck–Ein Beitrag zum Verständnis früher monumentaler Kultbauten Mitteleuropas. In K. Schmotz (Ed.), *Vorträge des 25. Niederbayerischen Archäologentages*. Verlag Marie Leidorf.

Best, E. (1921). Polynesian mnemonics: Notes on the use of the Quipus in Polynesia in former times; also some account of the introduction of the art of writing. *New Zealand Journal of Science and Technology, 4*, 67–74.

Biederman, I., & Vessel, E. A. (2006). Perceptual pleasure and the brain: A novel theory explains why the brain craves information and seeks it through the senses. *American Scientist, 94*, 247–253.

Bigelow, A. E., & Dugas, K. (2009). Relations among preschool children's understanding of visual perspective taking, false belief, and lying. *Journal of Cognition and Development, 9*, 411–433.

Bingham, P. M. (1999). Human uniqueness: A general theory. *Quarterly Review of Biology, 74*, 133–169.

Biron, B. (2019, July 10). Beauty has blown up to be a $532 billion industry—and analysts say that these 4 trends will make it even bigger. *Business Insider Australia.*

Bischof, N. (1985). *Das Rätzel Ödipus [The Oedipus riddle].* Piper.

Bischof-Köhler, D. (1985). Zur Phylogenese menschlicher Motivation [On the phylogeny of human motivation]. In L. H. Eckensberger & E. D. Lantermann (Eds.), *Emotion und Reflexivität.* Urban & Schwarzenberg.

Blom, T., Feuerriegel, D., Johnson, P., Bode, S., & Hogendoorn, H. (2020). Predictions drive neural representations of visual events ahead of incoming sensory information. *Proceedings of the National Academy of Sciences, 117,* 7510–7515.

Bloom, P. (2004, May 11). *Natural-born dualists: A talk with Paul Bloom.* Edge.org.

Blust, R. (1996). Austronesian culture history: The window of language. *Transactions of the American Philosophical Society, 86,* 28–35.

Boeckle, M., Schiestl, M., Frohnwieser, A., Gruber, R., Miller, R., Suddendorf, T., . . . Clayton, N. (2020). New Caledonian crows plan for specific future tool use. *Proceedings of the Royal Society B, 287,* 20201490.

Boesch, C., & Boesch, H. (1984). Mental map in wild chimpanzees: An analysis of hammer transports for nut cracking. *Primates, 25,* 160–170.

Boesch, C., Bombjaková, D., Meier, A., & Mundry, R. (2019). Learning curves and teaching when acquiring nut-cracking in humans and chimpanzees. *Scientific Reports, 9,* 1515.

Boltz, W. G. (2000). The invention of writing in China. *Oriens Extremus, 42,* 1–17.

Bonmatí, A., Gómez-Olivencia, A., Arsuaga, J.-L., Carretero, J. M., Gracia, A., Martínez, I., . . . Carbonell, E. (2010). Middle Pleistocene lower back and pelvis from an aged human individual from the Sima de los Huesos site, Spain. *Proceedings of the National Academy of Sciences, 107,* 18386–18391.

Bonta, M., Gosford, R., Eussen, D., Ferguson, N., Loveless, E., & Witwer, M. (2017). Intentional fire-spreading by "Firehawk" raptors in Northern Australia. *Journal of Ethnobiology, 37,* 700–718.

Bordo, M. D., & Schwartz, A. J. (1984). *A retrospective on the classical gold standard, 1821–1931.* University of Chicago Press.

Borges, J. L. (1967). *A personal anthology.* Grove Press.

Borrelle, S. B., Ringma, J., Law, K. L., Monnahan, C. C., Lebreton, L., McGivern, A., . . . Rochman, C. M. (2020). Predicted growth in plastic waste exceeds efforts to mitigate plastic pollution. *Science, 369,* 1515–1518.

Bostrom, N. (2002). Existential risks: Analyzing human extinction scenarios and related hazards. *Journal of Evolution and Technology, 9,* 1–31.

Boswell, J. (1791). *The life of Samuel Johnson, LL.D.* Henry Baldwin.

Boyd, R., Richerson, P. J., & Henrich, J. (2011). The cultural niche: Why social learning is essential for human adaptation. *Proceedings of the National Academy of Sciences, 108,* 10918–10925.

Boyer, P. (2008). Evolutionary economics of mental time travel? *Trends in Cognitive Sciences, 12,* 219–224.

Boyer, P. (2020). Why divination? Evolved psychology and strategic interaction in the production of truth. *Current Anthropology, 61,* 100–123.

Bradley, J., & Yanyuwa families. (2010). *Singing saltwater country: Journey to the songlines of Carpentaria.* Allen & Unwin.

Bramble, D. M., & Lieberman, D. E. (2004). Endurance running and the evolution of *Homo. Nature, 432*, 345–352.

Brand, S. (n.d.). *About Long Now: The clock and library projects*. Long Now Foundation.

Bräuer, J., & Call, J. (2015). Apes produce tools for future use. *American Journal of Primatology, 77*, 254–263.

Brezina, T., Tekin, E., & Topalli, V. (2009). "Might not be a tomorrow": A multimethods approach to anticipated early death and youth crime. *Criminology, 47*, 1091–1129.

Bricker, H. M., Aveni, A. F., & Bricker, V. R. (2001). Ancient Maya documents concerning the movements of Mars. *Proceedings of the National Academy of Sciences, 98*, 2107–2110.

Brinums, M., Imuta, K., & Suddendorf, T. (2018). Practicing for the future: Deliberate practice in early childhood. *Child Development, 89*, 2051–2058.

Brinums, M., Redshaw, J., Nielsen, M., Suddendorf, T., & Imuta, K. (2021). Young children's capacity to seek information in preparation for a future event. *Cognitive Development, 58*, 101015.

Bristow, W. (2017). Enlightenment. In E. N. Zalta (Ed.), *Stanford encyclopedia of philosophy*.

Britton, J. P. (1989). An early function for eclipse magnitudes in Babylonian astronomy. *Centaurus, 32*, 1–52.

Brooks, A. S., Yellen, J. E., Potts, R., Behrensmeyer, A. K., Deino, A. L., Leslie, D. E., . . . Clark, J. B. (2018). Long-distance stone transport and pigment use in the earliest Middle Stone Age. *Science, 360*, 90–94.

Brooks, R. C. (2011). *Sex, genes & rock' n' roll: How evolution has shaped the modern world*. ReadHowYouWant.com.

Brosnan, S. F., & de Waal, F. B. M. (2003). Monkeys reject unequal pay. *Nature, 425*, 297–299.

Brown, F., Harris, J., Leakey, R., & Walker, A. (1985). Early *Homo erectus* skeleton from West Lake Turkana, Kenya. *Nature, 316*, 788–792.

Browne, J. (2013). Wallace and Darwin. *Current Biology, 23*, R1071–R1072.

Bruce, C., & Salkeld, A. (Director). (2000). The love of money [Television series episode]. In C. Bruce, A. Salkeld, & B. Want (Executive producers), *The Road to Riches*. BBC Two.

Bryce, T. (2006). The "Eternal Treaty" from the Hittite perspective. *British Museum Studies in Ancient Egypt and Sudan*, 6–11.

Bryson, B. (2004). *A short history of nearly everything*. Broadway.

Buckner, R. L., & Carroll, D. C. (2007). Self-projection and the brain. *Trends in Cognitive Sciences, 11*, 49–57.

Buckner, R. L., & DiNicola, L. M. (2019). The brain's default network: Updated anatomy, physiology and evolving insights. *Nature Reviews Neuroscience, 20*, 593–608.

Buckner, W., & Glowacki, L. (2019). Reasons to strike first. *Behavioral and Brain Sciences, 42*, 17–18.

Buehler, R., Griffin, D., & Peetz, J. (2010). The planning fallacy: Cognitive, motivational, and social origins. *Advances in Experimental Social Psychology, 43*, 1–62.

Bulley, A., Henry, J., & Suddendorf, T. (2016). Prospection and the present moment: The role of episodic foresight in intertemporal choices between immediate and delayed rewards. *Review of General Psychology, 20*, 29–47.

Bulley, A., & Irish, M. (2018). The functions of prospection—Variations in health and disease. *Frontiers in Psychology, 9*, 2328.

Bulley, A., McCarthy, T., Gilbert, S. J., Suddendorf, T., & Redshaw, J. (2020). Children devise and selectively use tools to offload cognition. *Current Biology, 30*, 3457–3464.

Bulley, A., Miloyan, B., Pepper, G. V., Gullo, M. J., Henry, J. D., & Suddendorf, T. (2019). Cuing both positive and negative episodic foresight reduces delay discounting but does not affect risk-taking. *Quarterly Journal of Experimental Psychology, 72*, 1998–2017.

Bulley, A., & Pepper, G. V. (2017). Cross-country relationships between life expectancy, intertemporal choice and age at first birth. *Evolution and Human Behavior, 38*, 652–658.

Bulley, A., Pepper, G., & Suddendorf, T. (2017). Using foresight to prioritise the present. *Behavioral and Brain Sciences, 40*, 15–16.

Bulley, A., Redshaw, J., & Suddendorf, T. (2020). The future-directed functions of the imagination: From prediction to metaforesight. In A. Abraham (Ed.), *The Cambridge handbook of the imagination*. Cambridge University Press.

Bulley, A., & Schacter, D. L. (2020). Deliberating trade-offs with the future. *Nature Human Behaviour, 4*, 238–247.

Bulliet, R. W. (2016). *The wheel: Inventions and reinventions*. Columbia University Press.

Buonomano, D. (2017). *Your brain is a time machine: The neuroscience and physics of time*. WW Norton & Company.

Burgess, N. (2014). The 2014 Nobel Prize in Physiology or Medicine: A spatial model for cognitive neuroscience. *Neuron, 84*, 1120–1125.

Burgess, P. W., Dumontheil, I., & Gilbert, S. J. (2007). The gateway hypothesis of rostral prefrontal cortex (area 10) function. *Trends in Cognitive Sciences, 11*, 290–298.

Busby Grant, J., & Suddendorf, T. (2005). Recalling yesterday and predicting tomorrow. *Cognitive Development, 20*, 362–372.

Busby Grant, J., & Suddendorf, T. (2009). Preschoolers begin to differentiate the times of events from throughout the lifespan. *European Journal of Developmental Psychology, 6*, 746–762.

Busby Grant, J., & Suddendorf, T. (2010). Young children's ability to distinguish past and future changes in physical and mental states. *British Journal of Developmental Psychology, 28*, 853–870.

Busby Grant, J., & Suddendorf, T. (2011). Production of temporal terms by 3-, 4-, and 5-year-old children. *Early Childhood Research Quarterly, 26*, 87–95.

Buss, D. M. (1999). *Evolutionary psychology: The new science of the mind*. Allyn and Bacon.

Butler, A. C. (2010). Repeated testing produces superior transfer of learning relative to repeated studying. *Journal of Experimental Psychology: Learning, Memory, and Cognition, 36*, 1118–1133.

Butler, D. L., Mattingley, J. B., Cunnington, R., & Suddendorf, T. (2012). Mirror, mirror on the wall, how does my brain recognize my image at all? *PLOS ONE, 7*, e31452.

Butler, D. L., Mattingley, J. B., Cunnington, R., & Suddendorf, T. (2013). Different neural processes accompany self-recognition in photographs across the lifespan: An ERP study using dizygotic twins. *PLOS ONE, 8*, e72586.

Butler, D. L., & Suddendorf, T. (2014). Reducing the neural search space for hominid cognition: What distinguishes human and great ape brains from those of small apes? *Psychonomic Bulletin and Review, 21*, 590–619.

Calvin, W. H. (1982). Did throwing stones shape hominid brain evolution? *Ethology and Sociobiology, 3*, 115–124.

Canadian Press. (2007, November 23). Toronto-area amnesiac helps neuroscientists. *CTV News*.

Capasso, L. (1998). 5300 years ago, the Ice Man used natural laxatives and antibiotics. *The Lancet, 352*, 1864.

Caro, T. M., & Hauser, M. D. (1992). Is there teaching in nonhuman animals? *Quarterly Review of Biology, 67*, 151–174.

Carr, N. (2008, July/August). Is Google making us stupid? *The Atlantic*.

Casey, B. J., Somerville, L. H., Gotlib, I. H., Ayduk, O., Franklin, N. T., Askren, M. K., ... Shoda, Y. (2011). Behavioral and neural correlates of delay of gratification 40 years later. *Proceedings of the National Academy of Sciences, 108*, 14998–15003.

Caza, J. S., O'Brien, B. M., Cassidy, K. S., Ziani-Bey, H. A., & Atance, C. M. (2021). Tomorrow will be different: Children's ability to incorporate an intervening event when thinking about the future. *Developmental Psychology, 57*, 376–385.

Ceballos, G., Ehrlich, P. R., & Dirzo, R. (2017). Biological annihilation via the ongoing sixth mass extinction signaled by vertebrate population losses and declines. *Proceedings of the National Academy of Sciences, 114*, E6089–E6096.

Chadwick, R. (1984). The origins of astronomy and astrology in Mesopotamia. *Archaeoastronomy, 7*, 89–95.

Chatwin, B. (1987). *The songlines*. Penguin.

Cheke, L. G., & Clayton, N. S. (2012). Eurasian jays (*Garrulus glandarius*) overcome their current desires to anticipate two distinct future needs and plan for them appropriately. *Biology Letters, 8*, 171–175.

Cheng, S., Werning, M., & Suddendorf, T. (2016). Dissociating memory traces and scenario construction in mental time travel. *Neuroscience and Biobehavioral Reviews, 60*, 82–89.

Chomsky, N. (1959). A review of B. F. Skinner's Verbal Behavior. *Language, 35*, 26–58.

Chou, H.-h. (1979). Chinese oracle bones. *Scientific American, 240*, 134–149.

Cicero. (1853). *On divination* (C. D. Yonge, Trans.). H. G. Bohn. (Original work published 44 BC)

Cicero. (1877). *On the nature of the gods* (C. D. Yonge, Trans.). Harper & Brothers. (Original work published 45 BC)

Cicero. (1923). *On divination* (W. A. Falconer, Trans.). Loeb Classical Library. (Original work published 44 BC)

Clark, A. (1996). *Being there: Putting brain, body, and world together again*. MIT Press.

Clark, A. (2013). Whatever next? Predictive brains, situated agents, and the future of cognitive science. *Behavioral and Brain Sciences, 36*, 181–204.

Clark, A. (2016). *Surfing uncertainty: Prediction, action, and the embodied mind*. Oxford University Press.

Clark, A., & Chalmers, D. (1998). The extended mind. *Analysis, 58*, 7–19.

Clark, I. A., Monk, A. M., & Maguire, E. A. (2020). Characterizing strategy use during the performance of hippocampal-dependent tasks. *Frontiers in Psychology, 11*, 2119.

Clarke, P. A. (2009). Australian Aboriginal ethnometeorology and seasonal calendars. *History and Anthropology, 20*, 79–106.

Clarkson, C., Jacobs, Z., Marwick, B., Fullagar, R., Wallis, L., Smith, M., . . . Pardoe, C. (2017). Human occupation of northern Australia by 65,000 years ago. *Nature, 547,* 306–310.

Clayton, N. S., & Dickinson, A. (1998). Episodic-like memory during cache recovery by scrub jays. *Nature, 395,* 272–278.

Clayton, N. S., Griffiths, D. P., Emery, N. J., & Dickinson, A. (2001). Elements of episodic-like memory in animals. *Philosophical Transactions of the Royal Society of London B: Biological Sciences, 356,* 1483–1491.

Clindaniel, J. (2019). *Toward a grammar of the Inka Khipu: Investigating the production of non-numerical signs.* (Unpublished doctoral dissertation). Harvard University.

Coffman, D. D. (1990). Effects of mental practice, physical practice, and knowledge of results on piano performance. *Journal of Research in Music Education, 38,* 187–196.

Collias, N. E., & Collias, E. C. (1984). *Nest building and bird behavior.* Princeton University Press.

Collier-Baker, E., & Suddendorf, T. (2006). Do chimpanzees (*Pan troglodytes*) and 2-year-old children (*Homo sapiens*) understand double invisible displacement? *Journal of Comparative Psychology, 120,* 89–97.

Conard, N. J., Malina, M., & Münzel, S. C. (2009). New flutes document the earliest musical tradition in southwestern Germany. *Nature, 460,* 737–740.

Conard, N. J., Serangeli, J., Bigga, G., & Rots, V. (2020). A 300,000-year-old throwing stick from Schöningen, northern Germany, documents the evolution of human hunting. *Nature Ecology & Evolution, 4,* 690–693.

Copernicus, N. (1976). *On the revolutions of the heavenly spheres* (A. M. Duncan, Trans.). Barnes & Noble Books. (Original work published 1543)

Coppens, Y. (1994). East side story: The origin of humankind. *Scientific American, 270,* 62–69.

Corballis, M. C. (2002). *From hand to mouth: The origins of language.* Princeton University Press.

Corballis, M. C. (2011). *The recursive mind: The origins of human language, thought, and civilization.* Princeton University Press.

Corballis, M. C. (2013a). Mental time travel: A case for evolutionary continuity. *Trends in Cognitive Sciences, 17,* 5–6.

Corballis, M. C. (2013b). The wandering rat: Response to Suddendorf. *Trends in Cognitive Sciences, 17,* 152.

Corballis, M. C. (2015). *The wandering mind: What the brain does when you're not looking.* University of Chicago Press.

Corballis, M. C. (2017). *The truth about language: What it is and where it came from.* University of Chicago Press.

Corbey, R., Jagich, A., Vaesen, K., & Collard, M. (2016). The Acheulean handaxe: More like a bird's song than a Beatles' tune? *Evolutionary Anthropology, 25,* 6–19.

Corcoran, A. W., Pezzulo, G., & Hohwy, J. (2020). From allostatic agents to counterfactual cognisers: Active inference, biological regulation, and the origins of cognition. *Biology & Philosophy, 35,* 32.

Craik, K. J. W. (1943). *The nature of explanation.* Cambridge University Press.

Crew, B. (2014, June 4). The legend of Old Tom and the gruesome "law of the tongue." *Scientific American*.

Cross, F. R., & Jackson, R. R. (2019). Portia's capacity to decide whether a detour is necessary. *Journal of Experimental Biology, 222*, jeb203463.

Crystal, J. D., & Suddendorf, T. (2019). Episodic memory in nonhuman animals? *Current Biology, 29*, R1291–R1295.

D'Argembeau, A. (2020). Zooming in and out on one's life: Autobiographical representations at multiple time scales. *Journal of Cognitive Neuroscience, 32*, 2037–2055.

D'Argembeau, A., Raffard, S., & Van der Linden, M. (2008). Remembering the past and imagining the future in schizophrenia. *Journal of Abnormal Psychology, 117*, 247–251.

D'Argembeau, A., & Van der Linden, M. (2004). Phenomenal characteristics associated with projecting oneself back into the past and forward into the future: Influence of valence and temporal distance. *Consciousness and Cognition, 13*, 844–858.

Darling, D. (2004). *The universal book of mathematics from abracadabra to Zeno's paradoxes*. John Wiley & Sons, Inc.

Darwin, C. (1871). *The descent of man, and selection in relation to sex*. John Murray.

Darwin, C. (1958). *The autobiography of Charles Darwin 1809–1882. With original omissions restored. Edited with Appendix and Notes by his grand-daughter Nora Barlow.* (N. Barlow, Ed.). Collins. (Original work published 1887)

Davidson, M. C., Amso, D., Anderson, L. C., & Diamond, A. (2006). Development of cognitive control and executive functions from 4 to 13 years: Evidence from manipulations of memory, inhibition, and task switching. *Neuropsychologia, 44*, 2037–2078.

Davies, N. B., Butchart, S. H. M., Burke, T. A., Chaline, N., & Stewart, I. R. K. (2003). Reed warblers guard against cuckoos and cuckoldry. *Animal Behaviour, 65*, 285–295.

Davies, S. (2021). Behavioral modernity in retrospect. *Topoi, 40*, 221–232.

Davis, J., Cullen, E., & Suddendorf, T. (2016). Understanding deliberate practice in preschool-aged children. *Quarterly Journal of Experimental Psychology, 69*, 361–380.

Davis, J., Redshaw, J., Suddendorf, T., Nielsen, M., Kennedy-Costantini, S., Oostenbroek, J., & Slaughter, V. (2021). Does neonatal imitation exist? Insights from a meta-analysis of 336 effect sizes. *Perspectives on Psychological Science, 16*, 1373–1397.

Davy, H. (1800). *Researches, chemical and philosophical; chiefly concerning nitrous oxide: Or dephlogisticated nitrous air, and its respiration*. J. Johnson.

Dawes, A. J., Keogh, R., Andrillon, T., & Pearson, J. (2020). A cognitive profile of multisensory imagery, memory and dreaming in aphantasia. *Scientific Reports, 10*, 10022.

Dawkins, R. (1986). *The blind watchmaker: Why the evidence of evolution reveals a universe without design*. Norton.

Dawkins, R., Krebs, J. R., Maynard Smith, J., & Holliday, R. (1979). Arms races between and within species. *Proceedings of the Royal Society of London B: Biological Sciences, 205*, 489–511.

De Jong, J. P., Von Hippel, E., Gault, F., Kuusisto, J., & Raasch, C. (2015). Market failure in the diffusion of consumer-developed innovations: Patterns in Finland. *Research Policy, 44*, 1856–1865.

De Vos, J. M., Joppa, L. N., Gittleman, J. L., Stephens, P. R., & Pimm, S. L. (2015). Estimating the normal background rate of species extinction. *Conservation Biology, 29*, 452–462.

Dean, L. G., Kendal, R. L., Schapiro, S. J., Thierry, B., & Laland, K. N. (2012). Identification of the social and cognitive processes underlying human cumulative culture. *Science, 335*, 1114–1118.

Dehaene, S. (2014). *Consciousness and the brain: Deciphering how the brain codes our thoughts.* Penguin.

Delsuc, F. (2003). Army ants trapped by their evolutionary history. *PLOS Biology, 1*, 155–156.

Dennell, R. (2011). An earlier Acheulian arrival in South Asia. *Science, 331*, 1532.

Dennett, D. (1999, August 15). *We earth neurons.* Tufts University website.

Dennett, D. (2009). Intentional systems theory. In A. Beckermann, B. P. McLaughlin, & S. Walter (Eds.), *The Oxford handbook of philosophy of mind.* Oxford University Press.

Derex, M., Bonnefon, J.-F., Boyd, R., & Mesoudi, A. (2019). Causal understanding is not necessary for the improvement of culturally evolving technology. *Nature Human Behaviour, 3*, 446–452.

d'Errico, F., Backwell, L., Villa, P., Degano, I., Lucejko, J. J., Bamford, M. K., . . . Beaumont, P. B. (2012). Early evidence of San material culture represented by organic artifacts from Border Cave, South Africa. *Proceedings of the National Academy of Sciences, 109*, 13214–13219.

Détroit, F., Mijares, A. S., Corny, J., Daver, G., Zanolli, C., Dizon, E., . . . Piper, P. J. (2019). A new species of Homo from the Late Pleistocene of the Philippines. *Nature, 568*, 181–186.

Di Marco, M., Venter, O., Possingham, H. P., & Watson, J. E. M. (2018). Changes in human footprint drive changes in species extinction risk. *Nature Communications, 9*, 4621.

Diamond, A., & Taylor, C. (1996). Development of an aspect of executive control: Development of the abilities to remember what I said and to "Do as I say, not as I do". *Developmental Psychobiology, 29*, 315–334.

Diamond, J. (1997). *Guns, germs and steel: A short history of everybody for the last 13,000 years.* Simon and Schuster.

Dickerson, K. L., Ainge, J. A., & Seed, A. M. (2018). The role of association in pre-schoolers' solutions to "spoon tests" of future planning. *Current Biology, 28*, 2309–2313.

Dickinson, A. (2012). Associative learning and animal cognition. *Philosophical Transactions of the Royal Society B: Biological Sciences, 367*, 2733–2742.

Diggins, F. W. (1999). The true history of the discovery of penicillin, with refutation of the misinformation in the literature. *British Journal of Biomedical Science, 56*, 83–93.

Doebel, S., Michaelson, L. E., & Munakata, Y. (2019). Good things come to those who wait: Delaying gratification likely does matter for later achievement (A commentary on Watts, Duncan, & Quan, 2018). *Psychological Science, 31*, 97–99.

Dohrn-van Rossum, G. (1996). *History of the hour: Clocks and modern temporal orders* (T. Dunlap, Trans.). University of Chicago Press. (Original work published 1992)

Donald, M. (1999). Preconditions for the evolution of protolanguages. In M. C. Corballis & S. E. G. Lea (Eds.), *The descent of mind: Psychological perspectives on hominid evolution.* Oxford University Press.

Donnachie, I. (2000). *Robert Owen: Owen of New Lanark and New Harmony.* Tuckwell Press.

Donovan, J. J., & Radosevich, D. J. (1999). A meta-analytic review of the distribution of practice effect: Now you see it, now you don't. *Journal of Applied Psychology, 84,* 795–805.

Drabble, J. (2015, October 4). Clemmons teenager at the pinnacle of sport stacking. *Winston-Salem Journal.*

Dragoi, G., & Tonegawa, S. (2011). Preplay of future place cell sequences by hippocampal cellular assemblies. *Nature, 469,* 397–401.

Dudai, Y., & Carruthers, M. (2005). The Janus face of Mnemosyne. *Nature, 434,* 567.

Dudgeon, C. L., Dunlop, R. A., & Noad, M. J. (2018). Modelling heterogeneity in detection probabilities in land and aerial abundance surveys in humpback whales (*Megaptera novaeangliae*). *Population Ecology, 60,* 371–387.

Dufour, V., & Sterck, E. H. M. (2008). Chimpanzees fail to plan in an exchange task but succeed in a tool-using procedure. *Behavioural Processes, 79,* 19–27.

Duncan, D. E. (2011). *The calendar* (4th ed.). Fourth Estate.

Dunsworth, H., & Buchanan, A. (2017, August 9). Sex makes babies. *Aeon.*

Eger II, E. I., Saidman, L. J., & Westhorpe, R. N. (2014). *The wondrous story of anesthesia.* Springer.

Eichenbaum, H. (2014). Time cells in the hippocampus: a new dimension for mapping memories. *Nature Reviews Neuroscience, 15,* 732–744.

Elster, J. (2000). *Ulysses unbound: Studies in rationality, precommitment, and constraints.* Cambridge University Press.

Engelmann, J. M., Clift, J. B., Herrmann, E., & Tomasello, M. (2017). Social disappointment explains chimpanzees' behaviour in the inequity aversion task. *Proceedings of the Royal Society B: Biological Sciences, 284,* 20171502.

Engelmann, J. M., Völter, C. J., O'Madagain, C., Proft, M., Haun, D. B., Rakoczy, H., & Herrmann, E. (2021). Chimpanzees consider alternative possibilities. *Current Biology, 31,* R1377–R1378.

Epstein, R. A., Patai, E. Z., Julian, J. B., & Spiers, H. J. (2017). The cognitive map in humans: spatial navigation and beyond. *Nature Neuroscience, 20,* 1504–1513.

Ericsson, K. A., Krampe, R. T., & Tesch-Römer, C. (1993). The role of deliberate practice in the acquisition of expert performance. *Psychological Review, 100,* 363–406.

Estrada, A., Garber, P. A., Rylands, A. B., Roos, C., Fernandez-Duque, E., Di Fiore, A., . . . Li, B. (2017). Impending extinction crisis of the world's primates: Why primates matter. *Science Advances, 3,* e1600946.

Eustache, F., Desgranges, B., & Messerli, P. (1996). Edouard Claparede and human memory. *Revue Neurologique, 152,* 602–610.

Evans, T. A., & Beran, M. J. (2007). Chimpanzees use self-distraction to cope with impulsivity. *Biology Letters, 3,* 599–602.

Falk, A., Kosse, F., & Pinger, P. (2020). Re-revisiting the marshmallow test: A direct comparison of studies by Shoda, Mischel, and Peake (1990) and Watts, Duncan, and Quan (2018). *Psychological Science, 31,* 100–104.

Falk, D. (2009). *Finding our tongues: Mothers, infants, and the origins of language.* Basic Books.

Feeney, W. E., Welbergen, J. A., & Langmore, N. E. (2012). The frontline of avian brood parasite–host coevolution. *Animal Behaviour, 84,* 3–12.

Ferguson, N. (2008). *The ascent of money: A financial history of the world.* Penguin.

Ferster, C. B., & Skinner, B. F. (1957). *Schedules of reinforcement*. Appleton-Century-Crofts.

Feynman, R. P. (1955). The value of science. *Engineering and Science, 19*, 13–15.

Finlayson, S., Finlayson, G., Guzman, F. G., & Finlayson, C. (2019). Neanderthals and the cult of the Sun Bird. *Quaternary Science Reviews, 217*, 217–224.

Fitzpatrick, S. M., & McKeon, S. (2020). Banking on stone money: Ancient antecedents to Bitcoin. *Economic Anthropology, 7*, 7–21.

Fivush, R. (2011). The development of autobiographical memory. *Annual Review of Psychology, 62*, 559–582.

Fivush, R., Haden, C. A., & Reese, E. (2006). Elaborating on elaborations: Role of maternal reminiscing style in cognitive and socioemotional development. *Child Development, 77*, 1568–1588.

Flannery, T. (1994). *The future eaters: An ecological history of the Australian lands and people*. Grove Press.

Fleming, S. M. (2021). *Know thyself*. Basic Books.

Fleming, S. M., & Dolan, R. J. (2012). The neural basis of metacognitive ability. *Philosophical Transactions of the Royal Society B: Biological Sciences, 367*, 1338–1349.

Foster, D. J., & Wilson, M. A. (2007). Hippocampal theta sequences. *Hippocampus, 17*, 1093–1099.

Fox, K. C., Dixon, M. L., Nijeboer, S., Girn, M., Floman, J. L., Lifshitz, M., . . . Christoff, K. (2016). Functional neuroanatomy of meditation: A review and meta-analysis of 78 functional neuroimaging investigations. *Neuroscience & Biobehavioral Reviews, 65*, 208–228.

Frank, R. H. (1988). *Passions within reason: The strategic role of the emotions*. WW Norton & Company.

Freeth, T., Higgon, D., Dacanalis, A., MacDonald, L., Georgakopoulou, M., & Wojcik, A. (2021). A model of the cosmos in the ancient Greek Antikythera mechanism. *Scientific Reports, 11*, 5821.

Friedman, W. J. (1990). *About time: Inventing the fourth dimension*. MIT Press.

Friston, K. (2010). The free-energy principle: A unified brain theory? *Nature Reviews Neuroscience, 11*, 127–138.

Fry, S. (1996). *Making history*. Random House.

Gaesser, B. (2020). Episodic mindreading: Mentalizing guided by scene construction of imagined and remembered events. *Cognition, 203*, 104325.

Gaffney, V. L., Fitch, S., Ramsey, E., Yorston, R., Ch'ng, E., Baldwin, E., . . . Sparrow, T. (2013). Time and a place: A luni-solar "time-reckoner" from 8th millennium BC Scotland. *Internet Archaeology, 34*, ia.34.31.

Galéra, C., Orriols, L., M'Bailara, K., Laborey, M., Contrand, B., Ribéreau-Gayon, R., . . . Fort, A. (2012). Mind wandering and driving: Responsibility case-control study. *BMJ, 345*, e8105.

Gallup, G. G. (1970). Chimpanzees: Self recognition. *Science, 167*, 86–87.

Gao, A. F., Keith, J. L., Gao, F. Q., Black, S. E., Moscovitch, M., & Rosenbaum, R. S. (2020). Neuropathology of a remarkable case of memory impairment informs human memory. *Neuropsychologia, 140*, 107342.

Garcia, J., Ervin, F. R., & Koelling, R. (1966). Learning with prolonged delay of reinforcement. *Psychonomic Science, 5*, 121–122.

Garrido, M. I., Teng, C. L. J., Taylor, J. A., Rowe, E. G., & Mattingley, J. B. (2016). Surprise responses in the human brain demonstrate statistical learning under high concurrent cognitive demand. *npj Science of Learning, 1*, 1–7.

Gautam, S., Bulley, A., von Hippel, W., & Suddendorf, T. (2017). Affective forecasting bias in preschool children. *Journal of Experimental Child Psychology, 159*, 175–184.

Gautam, S., Suddendorf, T., Henry, J. D., & Redshaw, J. (2019). A taxonomy of mental time travel and counterfactual thought: Insights from cognitive development. *Behavioural Brain Research, 374*, 112108.

Gautam, S., Suddendorf, T., & Redshaw, J. (2021). When can young children reason about an exclusive disjunction? *Cognition, 207*, 104507.

Gautam, S., Suddendorf, T., & Redshaw, J. (unpublished). What could have happened: Do young children experience regret and relief?

Gebhard, R., & Krause, R. (2020). Critical comments on the find complex of the so-called Nebra Sky Disk. *Archäologische Informationen, 43*, 325–346.

Gergeley, G., Nadasdy, Z., Csibra, G., & Biro, S. (1995). Taking the intentional stance at 12 months of age. *Cognition, 56*, 165–193.

Ghetti, S., & Coughlin, C. (2018). Stuck in the present? Constraints on children's episodic prospection. *Trends in Cognitive Sciences, 22*, 846–850.

Ghezzi, I., & Ruggles, C. (2007). Chankillo: A 2300-year-old solar observatory in coastal Peru. *Science, 315*, 1239–1243.

Gigerenzer, G. (1998). Ecological intelligence: An adaptation for frequencies. In D. D. Cummins & C. Allen (Eds.), *The evolution of mind*. Oxford University Press.

Gilbert, D. T. (2006). *Stumbling on happiness*. A.A. Knopf.

Gilbert, D. T., & Wilson, T. D. (2007). Prospection: Experiencing the future. *Science, 317*, 1351–1354.

Gilbert, D. T., Wilson, T. D., Pinel, E. C., Blumberg, S. J., & Wheatley, T. P. (1998). Immune neglect: A source of durability bias in affective forecasting. *Journal of Personality and Social Psychology, 75*, 617–638.

Gillman, M. A. (2019). Mini-review: A brief history of nitrous oxide (N_2O) use in neuropsychiatry. *Current Drug Research Reviews, 11*, 12–20.

Gladwell, M. (2008). *Outliers: The story of success*. Little, Brown.

Gleick, J. (2004). *Isaac Newton*. Vintage.

Glimcher, P. W. (2011). Understanding dopamine and reinforcement learning: The dopamine reward prediction error hypothesis. *Proceedings of the National Academy of Sciences, 108*, 15647–15654.

Gobet, F., Lane, P. C., Croker, S., Cheng, P. C., Jones, G., Oliver, I., & Pine, J. M. (2001). Chunking mechanisms in human learning. *Trends in Cognitive Sciences, 5*, 236–243.

Godfrey-Smith, P. (2016). *Other minds: The octopus and the evolution of intelligent life*. William Collins.

Golchert, J., Smallwood, J., Jefferies, E., Seli, P., Huntenburg, J. M., Liem, F., . . . Villringer, A. (2017). Individual variation in intentionality in the mind-wandering state is reflected in the integration of the default-mode, fronto-parietal, and limbic networks. *Neuroimage, 146*, 226–235.

Goldstone, J. A. (2000). The rise of the west—or not? A revision to socio-economic history. *Sociological Theory, 18*, 175–194.

Goodall, J. (1986). *The Chimpanzees of Gombe: Patterns of behaviour*. Harvard University Press.

Goren-Inbar, N., Alperson, N., Kislev, M. E., Simchoni, O., Melamed, Y., Ben-Nun, A., & Werker, E. (2004). Evidence of hominin control of fire at Gesher Benot Ya'aqov, Israel. *Science, 304*, 725–727.

Goren-Inbar, N., Lister, A., Werker, E., & Chech, M. (1994). A butchered elephant skull and associated artifacts from the Acheulian site of Gesher Benot Ya'aqov, Israel. *Paléorient, 20*, 99–112.

Gould, S. J. (1992). *Ever since Darwin: Reflections in natural history*. WW Norton & Company.

Gowlett, J. A. J. (2016). The discovery of fire by humans: A long and convoluted process. *Philosophical Transactions of the Royal Society B: Biological Sciences, 371*, 20150164.

Green, A. (2016). *"Oh excellent air bag": Under the influence of nitrous oxide 1799–1920*. Public Domain Review Press.

Green, D., Billy, J., & Tapim, A. (2010). Indigenous Australians' knowledge of weather and climate. *Climatic Change, 100*, 337–354.

Green, R. E., Krause, J., Briggs, A. W., Maricic, T., Stenzel, U., Kircher, M., . . . Paaebo, S. (2010). A draft sequence of the Neandertal genome. *Science, 328*, 710–722.

Greenberg, J., Solomon, S., & Pyszczynski, T. (1997). Terror management theory of self-esteem and cultural worldviews: Empirical assessments and conceptual refinements. *Advances in Experimental Social Psychology, 29*, 61–139.

Gribbin, J., & Gribbin, M. (2017). *Out of the shadow of a giant: Hooke, Halley, and the birth of science*. Yale University Press.

Gross, S. R., O'Brien, B., Hu, C., & Kennedy, E. H. (2014). Rate of false conviction of criminal defendants who are sentenced to death. *Proceedings of the National Academy of Sciences 111*, 7230–7235.

Gruber, R., Schiest, M., Boeckle, M., Frohnwieser, A., Miller, R., Gray, R. D., . . . Taylor, A. H. (2019). New Caledonian crows use mental representations to solve metatool problems. *Current Biology, 29*, 686–692.

Grun, R., Stringer, C., McDermott, F., Nathan, R., Porat, N., Robertson, S., . . . McCulloch, M. (2005). U-series and ESR analyses of bones and teeth relating to the human burials from Skhul. *Journal of Human Evolution, 49*, 316–334.

Guilford, J. P. (1967). *The nature of human intelligence*. McGraw-Hill.

Gurnett, D., Kurth, W., Burlaga, L., & Ness, N. (2013). In situ observations of interstellar plasma with Voyager 1. *Science, 341*, 1489–1492.

Gwinner, E. (2003). Circannual rhythms in birds. *Current Opinion in Neurobiology, 13*, 770–778.

Hafner, J. W., Sturgell, J. L., Matlock, D. L., Bockewitz, E. G., & Barker, L. T. (2012). "Stayin' alive": A novel mental metronome to maintain compression rates in simulated cardiac arrests. *Journal of Emergency Medicine, 43*, e373–377.

Hafting, T., Fyhn, M., Molden, S., Moser, M.-B., & Moser, E. I. (2005). Microstructure of a spatial map in the entorhinal cortex. *Nature, 436*, 801–806.

Halford, G. S., Wilson, W. H., & Phillips, S. (1998). Processing capacity defined by relational complexity: Implications for comparative, developmental and cognitive psychology. *Behavioral and Brain Sciences, 21*, 803–864.

Halley, E. (1693). An estimate of the degrees of the mortality of mankind; drawn from curious tables of the births and funerals at the city of Breslaw; with an attempt to ascertain the price of annuities upon lives. *Philosophical Transactions of the Royal Society of London, 17*, 596–610.

Hallos, J. (2005). "15 minutes of fame": Exploring the temporal dimension of Middle Plaistocene lithic technology. *Journal of Human Evolution, 49*, 155–179.

Hampton, R. (2019). Parallel overinterpretation of behavior of apes and corvids. *Learning & Behavior, 47*, 105–106.

Hanoch, Y., Rolison, J., & Freund, A. M. (2019). Reaping the benefits and avoiding the risks: Unrealistic optimism in the health domain. *Risk Analysis, 39*, 792–804.

Harari, Y. N. (2014). *Sapiens: A brief history of humankind*. Random House.

Hardy, B. L., Moncel, M. H., Kerfant, C., Lebon, M., Bellot-Gurlet, L., & Mélard, N. (2020). Direct evidence of Neanderthal fibre technology and its cognitive and behavioral implications. *Scientific Reports, 10*, 4889.

Haridas, R. P. (2013). Horace Wells' demonstration of nitrous oxide in Boston. *Anesthesiology, 119*, 1014–1022.

Harmand, S., Lewis, J. E., Feibel, C. S., Lepre, C. J., Prat, S., Lenoble, A., . . . Roche, H. (2015). 3.3-million-year-old stone tools from Lomekwi 3, West Turkana, Kenya. *Nature, 521*, 310–315.

Harre, N. (2018). *Psychology for a better world*. Department of Psychology. Auckland University Press.

Harris, P. L., German, T., & Mills, P. (1996). Children's use of counterfactual thinking in causal reasoning. *Cognition, 61*, 233–259.

Harvati, K., Röding, C., Bosman, A. M., Karakostis, F. A., Grün, R., Stringer, C., . . . Kouloukoussa, M. (2019). Apidima Cave fossils provide earliest evidence of *Homo sapiens* in Eurasia. *Nature, 571*, 500–504.

Hassabis, D., Kumaran, D., Vann, S. D., & Maguire, E. A. (2007). Patients with hippocampal amnesia cannot imagine new experiences. *Proceedings of the National Academy of Sciences, 104*, 1726–1731.

Hauser, M. D., Chomsky, N., & Fitch, W. T. (2002). The faculty of language: What is it, who has it, and how did it evolve? *Science, 298*, 1569–1579.

Helmholtz, H. v. (1925). *Treatise on physiological optics. Vol. 3: The perceptions of vision* (J. P. C. Southall, Trans.). Optical Society of America. (Original work published 1866)

Henrich, J. (2015). *The secret of our success*. Princeton University Press.

Henry, J. D., Addis, D. R., Suddendorf, T., & Rendell, P. G. (2016). Introduction to the Special Issue: Prospection difficulties in clinical populations. *British Journal of Clinical Psychology, 55*, 1–3.

Henshilwood, C., d'Errico, F., Vanhaeren, M., van Niekerk, K., & Jacobs, Z. (2004). Middle Stone Age shell beads from South Africa. *Science, 304*, 404.

Henshilwood, C. S., d'Errico, F., & Watts, I. (2009). Engraved ochres from the Middle Stone Age levels at Blombos Cave, South Africa. *Journal of Human Evolution, 57*, 27–47.

Henshilwood, C. S., d'Errico, F., van Niekerk, K. L., Dayet, L., Queffelec, A., & Pollarolo, L. (2018). An abstract drawing from the 73,000-year-old levels at Blombos Cave, South Africa. *Nature, 562*, 115–118.

Hero of Alexandria. (1851). *Pneumatica* (J. W. Greenwood, Trans.). Taylor, Walton & Maberly. (Original work published ca. 62 AD)

Herschel, J. (1830). *A preliminary discourse on the study of natural philosophy*. Longman Press.

Hershfield, H. E., John, E. M., & Reiff, J. S. (2018). Using vividness interventions to improve financial decision making. *Policy Insights from the Behavioral and Brain Sciences, 5*, 209–215.

Hershkovitz, I., Weber, G. W., Quam, R., Duval, M., Grün, R., Kinsley, L., . . . Weinstein-Evron, M. (2018). The earliest modern humans outside Africa. *Science, 359*, 456.

Hesslow, G. (2002). Conscious thought as simulation of behaviour and perception. *Trends in Cognitive Sciences, 6*, 242–247.

Hettiarachchi, S. (2018). Establishing the Indian Ocean Tsunami Warning and Mitigation System for human and environmental security. *Procedia Engineering, 212*, 1339–1346.

Heyes, C. (2016). Imitation: Not in our genes. *Current Biology, 26*, R412–R414.

Heyes, C. (2018). *Cognitive gadgets*. Harvard University Press.

Hiltzik, M. (1999). *Dealers of lightning: Xerox parc and the dawn of the computer age*. Harper Business.

Hobbes, T. (1651). *Leviathan*. Andrew Crooke.

Hoeffel, J. C. (2005). Prosecutorial discretion at the core: The good prosecutor meets Brady. *Dickinson Law Review, 109*, 1133–1154.

Hoerl, C., & McCormack, T. (2019). Thinking in and about time: A dual systems perspective on temporal cognition. *Behavioral and Brain Sciences, 42*.

Hoffmann, D. L., Standish, C. D., García-Diez, M., Pettitt, P. B., Milton, J. A., Zilhão, J., . . . Pike, A. W. G. (2018). U-Th dating of carbonate crusts reveals Neandertal origin of Iberian cave art. *Science, 359*, 912–915.

Hofstadter, D. R. (1979). *Gödel, Escher, Bach*. Basic Books.

Holden, C. (2005). Time's up on time travel. *Science, 308*, 1110.

Holloway, R. L. (2008). The human brain evolving: A personal retrospective. *Annual Review of Anthropology, 37*, 1–19.

Holmes, R. (2008). *The age of wonder: How the Romantic generation discovered the beauty and terror of science*. Harper Press.

Hopkin, M. (2004). Sumatran quake sped up Earth's rotation. *Nature*, 10.1038.

Hoppitt, W. J. E., Brown, G. R., Kendal, R., Rendell, L., Thornton, A., Webster, M. M., & Laland, K. N. (2008). Lessons from animal teaching. *Trends in Ecology & Evolution, 23*, 486–493.

Horley, P. (2011). Lunar calendar in rongorongo texts and rock art of Easter Island. *Journal de la Société des Océanistes*, 17–38.

Horner, V., & Whiten, A. (2005). Causal knowledge and imitation/emulation switching in chimpanzees (*Pan troglodytes*) and children (*Homo sapiens*). *Animal Cognition, 8*, 164–181.

Hublin, J.-J., Ben-Ncer, A., Bailey, S. E., Freidline, S. E., Neubauer, S., Skinner, M. M., . . . Gunz, P. (2017). New fossils from Jebel Irhoud, Morocco and the pan-African origin of *Homo sapiens*. *Nature, 546*, 289–292.

Hudson, J. A. (2006). The development of future time concepts through mother-child conversation. *Merrill Palmer Quarterly: Journal of Developmental Psychology, 52*, 70–95.

Hume, D. (1739). *A treatise of human nature*. John Noon.

Humphrey, N. (2000, July). *Great expectations: The evolutionary psychology of faith healing and the placebo effect.* Paper presented at the XXVII International Congress of Psychology, Stockholm, Sweden.

Hunger, H., & Steele, J. (2018). *The Babylonian astronomical compendium MUL.APIN.* Routledge.

Ikeya, N. (2015). Maritime transport of obsidian in Japan during the Upper Paleolithic. In Y. Kaifu (Ed.), *Emergence and diversity in modern human behaviour in Paleolithic Asia.* Texas A&M University Press.

Imamoglu, F., Kahnt, T., Koch, C., & Haynes, J.-D. (2012). Changes in functional connectivity support conscious object recognition. *Neuroimage, 63,* 1909–1917.

Imamoglu, F., Koch, C., & Haynes, J.-D. (2013). MoonBase: Generating a database of two-tone Mooney images. *Journal of Vision, 13,* 50.

Inoue, S., & Matsuzawa, T. (2007). Working memory of numerals in chimpanzees. *Current Biology, 17,* R1004–R1005.

Irish, M., Goldberg, Z.-l., Alaeddin, S., O'Callaghan, C., & Andrews-Hanna, J. R. (2019). Age-related changes in the temporal focus and self-referential content of spontaneous cognition during periods of low cognitive demand. *Psychological Research, 83,* 747–760.

Irish, M., Hodges, J. R., & Piguet, O. (2013). Episodic future thinking is impaired in the behavioural variant of frontotemporal dementia. *Cortex, 49,* 2377–2388.

Irish, M., Piguet, O., & Hodges, J. R. (2012). Self-projection and the default network in frontotemporal dementia. *Nature Reviews Neurology, 8,* 152–161.

Irish, M., & Piolino, P. (2016). Impaired capacity for prospection in the dementias—Theoretical and clinical implications. *British Journal of Clinical Psychology, 55,* 49–68.

Jacobsen, L. E. (1983). Use of knotted string accounting records in old Hawaii and ancient China. *Accounting Historians Journal, 10,* 53–61.

James, W. (1890). *The principles of psychology.* Macmillan.

Janik, V. M. (2015). Play in dolphins. *Current Biology, 25,* R7–R8.

Janmaat, K. R., Boesch, C., Byrne, R., Chapman, C. A., Goné Bi, Z. B., Head, J. S., . . . Polansky, L. (2016). Spatio-temporal complexity of chimpanzee food: How cognitive adaptations can counteract the ephemeral nature of ripe fruit. *American Journal of Primatology, 78,* 626–645.

Jerison, H. J. (1973). *The evolution of the brain and intelligence.* Academic Press.

Johnson, A., & Redish, A. D. (2007). Neural ensembles in CA3 transiently encode paths forward of the animal at a decision point. *Journal of Neuroscience, 27,* 12176–12189.

Johnson, S. (2021, April 27). How humanity gave itself an extra life. *New York Times Magazine.*

Jones, E. M. (1985). *"Where is everybody." An account of Fermi's question.* Los Alamos National Lab.

Kabadayi, C., & Osvath, M. (2017). Ravens parallel great apes in flexible planning for tool-use and bartering. *Science, 357,* 202–204.

Kahneman, D. (2011). *Thinking: Fast and slow.* Farrar, Straus & Giroux.

Kaku, M. (2021, December 15). *Albert Einstein.* Encyclopedia Britannica.

Kawai, N., & Matsuzawa, T. (2000). Numerical memory span in a chimpanzee. *Nature, 403,* 39–40.

Keeley, L. H. (1996). *War before civilization*. Oxford University Press.

Kelly, L. (2016). *The memory code*. Simon and Schuster.

Kendal, R. L. (2019). Explaining human technology. *Nature Human Behaviour, 3*, 422–423.

Ker, J. (2010). Nundinae: The culture of the Roman week. *Phoenix, 64*, 360–385.

Keven, N., & Akins, K. A. (2017). Neonatal imitation in context: Sensorimotor development in the perinatal period. *Behavioral and Brain Sciences, 40*, 1–58.

Kidd, C., Palmeri, H., & Aslin, R. N. (2013). Rational snacking: Young children's decision-making on the marshmallow task is moderated by beliefs about environmental reliability. *Cognition, 126*, 109–114.

Kilgannon, C. (2020, January 1). Don't drink and drive, Republican leader said. Then he was arrested. *New York Times*.

King, L. W. (1910). *Codex Hammurabi (King translation)*. Yale Law School.

Klein, G. (2007). Performing a project premortem. *Harvard Business Review, 85*, 18–19.

Klein, S. B., Cosmides, L., Gangi, C. E., Jackson, B., Tooby, J., & Costabile, K. A. (2009). Evolution and episodic memory. *Social Cognition, 27*, 283–319.

Klein, S. B., Loftus, J., & Kihlstrom, J. F. (2002). Memory and temporal experience: The effects of episodic memory loss on an amnesic patient's ability to remember the past and imagine the future. *Social Cognition, 20*, 353–379.

Koch, C. (2016). How the computer beat the Go player. *Scientific American Mind, 27*, 20–23.

Koch, P. L., & Barnosky, A. D. (2006). Late Quaternary extinctions: State of the debate. *Annual Review of Ecology, Evolution, and Systematics, 37*, 215–250.

Kolb, B. (2019, December 24). Drive safely this holiday season. *Daily Messenger*.

Kolbert, E. (2014). *The sixth extinction: An unnatural history*. A&C Black.

Kondo, N., Sekijima, T., Kondo, J., Takamatsu, N., Tohya, K., & Ohtsu, T. (2006). Circannual control of hibernation by HP complex in the brain. *Cell, 125*, 161–172.

Kooij, D. T., Kanfer, R., Betts, M., & Rudolph, C. W. (2018). Future time perspective: A systematic review and meta-analysis. *Journal of Applied Psychology, 103*, 867–893.

Kopp, L., Atance, C. M., & Pearce, S. (2017). "Things aren't so bad!": Preschoolers overpredict the emotional intensity of negative outcomes. *British Journal of Developmental Psychology, 35*, 623–627.

Košťál, J., Klicperová-Baker, M., Lukavská, K., & Lukavský, J. (2015). Short version of the Zimbardo Time Perspective Inventory (ZTPI–short) with and without the Future-Negative scale, verified on nationally representative samples. *Time & Society, 25*, 169–192.

Kozowyk, P. R. B., Soressi, M., Pomstra, D., & Langejans, G. H. J. (2017). Experimental methods for the Palaeolithic dry distillation of birch bark: Implications for the origin and development of Neandertal adhesive technology. *Scientific Reports, 7*, 8033.

Kramer, S. N. (1949). Schooldays: A Sumerian composition relating to the education of a scribe. *Journal of the American Oriental Society, 69*, 199–215.

Krupenye, C., Kano, F., Hirata, S., Call, J., & Tomasello, M. (2016). Great apes anticipate that other individuals will act according to false beliefs. *Science, 354*, 110–114.

Krznaric, R. (2020). *The good ancestor: How to think long term in a short-term world*. WH Allen.

Kuczaj, S. A., & Walker, R. T. (2006). How do dolphins solve problems? In E. A. Wasserman & T. R. Zentall (Eds.), *Comparative cognition: Experimental explorations of animal intelligence*. Oxford University Press.

Kuhn, D. (2012). The development of causal reasoning. *Wiley Interdisciplinary Reviews: Cognitive Science, 3*, 327–335.

Kulke, L., & Hinrichs, M. A. B. (2021). Implicit Theory of Mind under realistic social circumstances measured with mobile eye-tracking. *Scientific Reports, 11*, 1215.

Kurlansky, M. (1997). *Cod: A biography of the fish that changed the world*. Walker & Co.

Lagattuta, K. H. (2014). Linking past, present, and future: Children's ability to connect mental states and emotions across time. *Child Development Perspectives, 8*, 90–95.

Lahr, M. M., Rivera, F., Power, R. K., Mounier, A., Copsey, B., Crivellaro, F., . . . Lawrence, J. (2016). Inter-group violence among early Holocene hunter-gatherers of West Turkana, Kenya. *Nature, 529*, 394–398.

Lambert, M. L., & Osvath, M. (2018). Comparing chimpanzees' preparatory responses to known and unknown future outcomes. *Biology Letters, 14*, 20180499.

Langley, M. C., Amano, N., Wedage, O., Deraniyagala, S., Pathmalal, M. M., Perera, N., . . . Roberts, P. (2020). Bows and arrows and complex symbolic displays 48,000 years ago in the South Asian tropics. *Science Advances, 6*, eaba3831.

Langley, M. C., & Suddendorf, T. (2020). Mobile containers in human cognitive evolution studies: Understudied and underrepresented. *Evolutionary Anthropology, 29*, 299–309.

Laozi (1989). *Tao Te Ching* (G-F. Feng & J. English, Trans.). Vintage Books. (Original work ca. fourth to sixth century BC)

Laplace, P.-S. (1951). *A philosophical essay on probabilities* (F. W. Truscott and F. L. Emory, Trans.). Dover Publications. (Original work published 1814)

Larena, M., McKenna, J., Sanchez-Quinto, F., Bernhardsson, C., Ebeo, C., Reyes, R., . . . Jakobsson, M. (2021). Philippine Ayta possess the highest level of Denisovan ancestry in the world. *Current Biology, 31*, 4219–4230.

Larsen, D. P., Butler, A. C., & Roediger III, H. L. (2009). Repeated testing improves long-term retention relative to repeated study: A randomised controlled trial. *Medical Education, 43*, 1174–1181.

Laukkonen, R. E., & Slagter, H. A. (2021). From many to (n)one: Meditation and the plasticity of the predictive mind. *Neuroscience & Biobehavioral Reviews, 128*, 199–217.

Laville, S., & Taylor, M. (2017, June 28). A million bottles a minute: World's plastic binge "as dangerous as climate change". *The Guardian*.

Law, K. L., & Thompson, R. C. (2014). Microplastics in the seas. *Science, 345*, 144.

Leahy, B. P., & Carey, S. E. (2020). The acquisition of modal concepts. *Trends in Cognitive Sciences, 24*, 65–78.

Leakey, R. E. F. (1969). Early *Homo sapiens* remains from Omo River region of south west Ethiopia. *Nature, 222*, 1132–1134.

Leder, D., Hermann, R., Hüls, M., Russo, G., Hoelzmann, P., Nielbock, R., . . . Terberger, T. (2021). A 51,000-year-old engraved bone reveals Neanderthals' capacity for symbolic behaviour. *Nature Ecology & Evolution, 9*, 1273–1282.

LeDoux, J. (2015). *Anxious: The modern mind in the age of anxiety*. Simon and Schuster.

Lee, G., & Kim, J. (2015). Obsidians from the Sinbuk archaeological site in Korea—Evidences for strait crossing and long-distance exchange of raw material in Paleolithic Age. *Journal of Archaeological Science: Reports, 2*, 458–466.

Lee, J. L. C., Nader, K., & Schiller, D. (2017). An update on memory reconsolidation updating. *Trends in Cognitive Sciences, 21,* 531–545.

Lee, W. S. C., & Carlson, S. M. (2015). Knowing when to be "rational": Flexible economic decision making and executive function in preschool children. *Child Development, 86,* 1434–1448.

Lehoux, D. (2007). *Astronomy, weather, and calendars in the ancient world: Parapegmata and related texts in classical and Near-Eastern societies.* Cambridge University Press.

Lempert, K. M., Steinglass, J. E., Pinto, A., Kable, J. W., & Simpson, H. B. (2019). Can delay discounting deliver on the promise of RDoC? *Psychological Medicine, 49,* 190–199.

Lepre, C. J., Roche, H., Kent, D. V., Harmand, S., Quinn, R. L., Brugal, J. P., . . . Feibel, C. S. (2011). An earlier origin for the Acheulian. *Nature, 477,* 82–85.

Lewis, J. E., & Harmand, S. (2016). An earlier origin for stone tool making: Implications for cognitive evolution and the transition to *Homo. Philosophical Transactions of the Royal Society B, 371,* 20150233.

Lewis, S. L., & Maslin, M. A. (2015). Defining the Anthropocene. *Nature, 519,* 171–180.

Lillard, A. S. (2017). Why do the children (pretend) play? *Trends in Cognitive Sciences, 21,* 826–834.

Liu, S., Brooks, N. B., & Spelke, E. S. (2019). Origins of the concepts cause, cost, and goal in prereaching infants. *Proceedings of the National Academy of Sciences, 116,* 17747–17752.

Loewenstein, G. (1987). Anticipation and the valuation of delayed consumption. *Economic Journal, 97,* 666–684.

Loftus, E. F. (2001). Imagining the past. *Psychologist, 14,* 584–587.

Lordkipanidze, D., Vekua, A., Ferring, R., Rightmire, G. P., Agustill, J., Kiladze, G., . . . Zollikofer, C. P. E. (2005). The earliest toothless hominin skull. *Nature, 434,* 717–718.

Lorenz, K., & Tinbergen, N. (1939). Taxis und Instinkthandlung in der Eirollbewegung der Graugans. *Zeitschrift fur Tierpsychologie, 2,* 1–29.

Louys, J., Braje, T. J., Chang, C.-H., Cosgrove, R., Fitzpatrick, S. M., Fujita, M., . . . O'Connor, S. (2021). No evidence for widespread island extinctions after Pleistocene hominin arrival. *Proceedings of the National Academy of Sciences, 118,* e2023005118.

Lu, X., Kelly, M. O., & Risko, E. F. (2020). Offloading information to an external store increases false recall. *Cognition, 205,* 104428.

Luna, B., Garver, K. E., Urban, T. A., Lazar, N. A., & Sweeney, J. A. (2004). Maturation of cognitive processes from late childhood to adulthood. *Child Development, 75,* 1357–1372.

Luria, A. R. (1973). *The working brain: An introduction to neuropsychology* (B. Haigh, Trans.). Basic Books.

Lyon, D. L., & Flavell, J. (1994). Young children's understanding of "remember" and "forget". *Child Development, 65,* 1357–1371.

Lyons, A. D., Henry, J. D., Rendell, P. G., Corballis, M. C., & Suddendorf, T. (2014). Episodic foresight and aging. *Psychology and Aging, 29,* 873–884.

Lyons, A. D., Henry, J. D., Rendell, P. G., Robinson, G., & Suddendorf, T. (2016). Episodic foresight and schizophrenia. *British Journal of Clinical Psychology, 55,* 107–122.

Lyons, A. D., Henry, J. D., Robinson, G., Rendell, P. G., & Suddendorf, T. (2019). Episodic foresight and stroke. *Neuropsychology, 33,* 93–102.

Mack, K. (2020). *The end of everything (astrophysically speaking).* Simon and Schuster.

Macnamara, B. N., Hambrick, D. Z., & Oswald, F. L. (2014). Deliberate practice and performance in music, games, sports, education, and professions: A meta-analysis. *Psychological Science, 25*, 1608–1618.

Macnamara, B. N., & Maitra, M. (2019). The role of deliberate practice in expert performance. *Royal Society Open Science, 6*, 190327.

Macnamara, B. N., Moreau, D., & Hambrick, D. Z. (2016). The relationship between deliberate practice and performance in sports: A meta-analysis. *Perspectives on Psychological Science, 11*, 333–350.

Macrobius. (2011). *Saturnalia* (R. A. Kaster, Trans.). Harvard University Press. (Original work published 431 AD)

Mahr, J. B., & Csibra, G. (2018). Why do we remember? The communicative function of episodic memory. *Behavioral and Brain Sciences, 41*, e1.

Maier, S. F., & Seligman, M. E. P. (2016). Learned helplessness at fifty: Insights from neuroscience. *Psychological Review, 123*, 349–367.

Majolo, B. (2019). Warfare in an evolutionary perspective. *Evolutionary Anthropology: Issues, News, and Reviews, 28*, 321–331.

Malville, J. M., Schild, R., Wendorf, F., & Brenmer, R. (2007). Astronomy of Nabta Playa. *African Skies, 11*, 2–7.

Malville, J. M., Wendorf, F., Mazar, A. A., & Schild, R. (1998). Megaliths and Neolithic astronomy in southern Egypt. *Nature, 392*, 488–491.

Marino, L. (2007). Cetacean brains: How aquatic are they? *The Anatomical Record: Advances in Integrative Anatomy and Evolutionary Biology, 290*, 694–700.

Marino, L., Connor, R. C., Fordyce, R. E., Herman, L. M., Hof, P. R., Lefebvre, L., . . . Whitehead, H. (2007). Cetaceans have complex brains for complex cognition. *PLOS Biology, 5*, 966–972.

Marshack, A. (1972). Cognitive aspects of Upper Paleolithic engraving. *Current Anthropology, 13*, 445–477.

Martin-Ordas, G. (2020). It is about time: Conceptual and experimental evaluation of the temporal cognitive mechanisms in mental time travel. *Wiley Interdisciplinary Reviews-Cognitive Science, 11*, e1530.

Maslow, A. H. (1943). A theory of human motivation. *Psychological Review, 50*, 370–396.

Matisoo-Smith, E. (2015). Ancient DNA and the human settlement of the Pacific: A review. *Journal of Human Evolution, 79*, 93–104.

Mazza, S., Gerbier, E., Gustin, M., Kasikci, Z., Koenig, O., Toppino, T., & Magnin, M. (2016). Relearn faster and retain longer: Along with practice, sleep makes perfect. *Psychological Science, 27*, 1321–1330.

McBrearty, S., & Brooks, A. S. (2000). The revolution that wasn't: A new interpretation of the origin of modern human behavior. *Journal of Human Evolution, 39*, 453–563.

McCormack, T., Feeney, A., & Beck, S. R. (2020). Regret and decision-making: A developmental perspective. *Current Directions in Psychological Science, 29*, 346–350.

McCormack, T., Ho, M., Gribben, C., O'Connor, E., & Hoerl, C. (2018). The development of counterfactual reasoning about doubly-determined events. *Cognitive Development, 45*, 1–9.

McDougall, I., Brown, F. H., & Fleagle, J. G. (2005). Stratigraphic placement and age of modern humans from Kibish, Ethiopia. *Nature, 433*, 733–736.

McNeill, J. R. (2001). *Something new under the sun: An environmental history of the twentieth-century world.* WW Norton & Company.

McRobbie, L. R. (2021, November 13). Party like it's 2269. *Boston Globe.*

McWethy, D. B., Whitlock, C., Wilmshurst, J. M., McGlone, M. S., & Li, X. (2009). Rapid deforestation of South Island, New Zealand, by early Polynesian fires. *The Holocene, 19,* 883–897.

Meltzoff, A. N., & Decety, J. (2003). What imitation tells us about social cognition: A rapprochement between developmental psychology and cognitive neuroscience. *Philosophical Transactions of the Royal Society B, 358,* 491–500.

Meltzoff, A. N., & Moore, M. K. (1977). Imitation of facial and manual gestures by human neonates. *Science, 198,* 75–78.

Meltzoff, A. N., & Moore, M. K. (1983). Newborn infants imitate adult facial gestures. *Child Development, 54,* 702–709.

Mercader, J., Barton, H., Gillespie, J., Harris, J., Kuhn, S., Tyler, R., & Boesch, C. (2007). 4,300-year-old chimpanzee sites and the origins of percussive stone technology. *Proceedings of the National Academy of Sciences 104,* 3043–3048.

Mesoudi, A. (2008). Foresight in cultural evolution. *Biology and Philosophy, 23,* 243–255.

Mesoudi, A. (2021). Cultural selection and biased transformation: Two dynamics of cultural evolution. *Philosophical Transactions of the Royal Society B, 376,* 20200053.

Mesoudi, A., Whiten, A., & Laland, K. N. (2006). Towards a unified science of cultural evolution. *Behavioral and Brain Sciences, 29,* 329–347.

Milks, A., Parker, D., & Pope, M. (2019). External ballistics of Pleistocene hand-thrown spears: Experimental performance data and implications for human evolution. *Scientific Reports, 9,* 820.

Miller, G. (2000). *The mating mind: How sexual choice shaped the evolution of human nature.* Anchor.

Miloyan, B., Bulley, A., & Suddendorf, T. (2019). Anxiety: Here and beyond. *Emotion Review, 11,* 39–49.

Miloyan, B., & McFarlane, K. A. (2019). The measurement of episodic foresight: A systematic review of assessment instruments. *Cortex, 117,* 351–370.

Miloyan, B., McFarlane, K. A., & Suddendorf, T. (2019). Measuring mental time travel: Is the hippocampus really critical for episodic memory and episodic foresight? *Cortex, 117,* 371–384.

Miloyan, B., Pachana, N. A., & Suddendorf, T. (2014). The future is here: A review of foresight systems in anxiety and depression. *Cognition and Emotion, 28,* 795–810.

Miloyan, B., & Suddendorf, T. (2015). Feelings of the future. *Trends in Cognitive Sciences, 19,* 196–200.

Mischel, W. (1974). Processes in delay of gratification. In L. Berkowitz (Ed.), *Advances in experimental social psychology.* Academic Press.

Mischel, W., Ayduk, O., Berman, M. G., Casey, B. J., Gotlib, I. H., Jonides, J., . . . Shoda, Y. (2011). "Willpower" over the life span: Decomposing self-regulation. *Social Cognitive and Affective Neuroscience, 6,* 252–256.

Mischel, W., Shoda, Y., & Rodriguez, M. I. (1989). Delay of gratification in children. *Science, 244,* 933–938.

Mitchell, A., Romano, G. H., Groisman, B., Yona, A., Dekel, E., Kupiec, M., . . . Pilpel, Y. (2009). Adaptive prediction of environmental changes by microorganisms. *Nature, 460,* 220–224.

Mody, S., & Carey, S. (2016). The emergence of reasoning by the disjunctive syllogism in early childhood. *Cognition, 154,* 40–48.

Mooneyham, B. W., & Schooler, J. W. (2013). The costs and benefits of mind-wandering: A review. *Canadian Journal of Experimental Psychology, 67,* 11–18.

More, T. (1516). *Utopia.*

Morgan, T. J., Uomini, N. T., Rendell, L. E., Chouinard-Thuly, L., Street, S. E., Lewis, H. M., . . . de la Torre, I. (2015). Experimental evidence for the co-evolution of hominin tool-making teaching and language. *Nature Communications, 6,* 1–8.

Moser, E. I., Moser, M.-B., & McNaughton, B. L. (2017). Spatial representation in the hippocampal formation: A history. *Nature Neuroscience, 20,* 1448–1464.

Muenzinger, K. F. (1938). Vicarious trial and error at a point of choice: I. A general survey of its relation to learning efficiency. *Pedagogical Seminary and Journal of Genetic Psychology, 53,* 75–86.

Mukerjee, M. (2003). Circles for space. *Scientific American, 289,* 32–34.

Mulcahy, N. J., & Call, J. (2006). Apes save tools for future use. *Science, 312,* 1038–1040.

Mulcahy, N. J., Call, J., & Dunbar, R. I. M. (2005). Gorillas (*Gorilla gorilla*) and orangutans (*Pongo pygmaeus*) encode relevant problem features in a tool-using task. *Journal of Comparative Psychology, 119,* 23–32.

Musgrave, S., Lonsdorf, E., Morgan, D., Prestipino, M., Bernstein-Kurtycz, L., Mundry, R., & Sanz, C. (2020). Teaching varies with task complexity in wild chimpanzees. *Proceedings of the National Academy of Sciences, 117,* 969–976.

Napper, I. E., Davies, B. F. R., Clifford, H., Elvin, S., Koldewey, H. J., Mayewski, P. A., . . . Thompson, R. C. (2020). Reaching new heights in plastic pollution—preliminary findings of microplastics on Mount Everest. *One Earth, 3,* 621–630.

NASA. (2003). *Agency contingency action plan (CAP) for space flight operations (SFO).*

Navasky, V. (1996, September 29). Tomorrow never knows. *New York Times Magazine.*

Nesse, R. (1998). Emotional disorders in evolutionary perspective. *British Journal of Medical Psychology, 71,* 397–415.

Newton, I. (1687). *Philosophiæ naturalis principia mathematica.* Royal Society.

Nielsen, J., Hedeholm, R. B., Heinemeier, J., Bushnell, P. G., Christiansen, J. S., Olsen, J., . . . Steffensen, J. F. (2016). Eye lens radiocarbon reveals centuries of longevity in the Greenland shark (*Somniosus microcephalus*). *Science, 353,* 702–704.

Nielsen, M., & Dissanayake, C. (2004). Pretend play, mirror self-recognition and imitation: A longitudinal investigation through the second year. *Infant Behavior and Development, 27,* 342–365.

Nielsen, M., Dissanayake, C., & Kashima, Y. (2003). A longitudinal investigation of self-other discrimination and the emergence of mirror self-recognition. *Infant Behavior & Development, 26,* 213–226.

Nielsen, M., Suddendorf, T., & Slaughter, V. (2006). Mirror self-recognition beyond the face. *Child Development, 77,* 176–185.

Nijhawan, R. (2008). Visual prediction: Psychophysics and neurophysiology of compensation for time delays. *Behavioral and Brain Sciences, 31,* 179–198.

Noad, M. J., Dunlop, R. A., Paton, D., & Cato, D. H. (2020). Absolute and relative abundance estimates of Australian east coast humpback whales (*Megaptera novaeangliae*). *Journal of Cetacean Research and Management, 3*, 243–252.

Noad, M. J., Kniest, E., & Dunlop, R. A. (2019). Boom to bust? Implications for the continued rapid growth of the eastern Australian humpback whale population despite recovery. *Population Ecology, 61*, 198–209.

Nobel, W., & Davidson, I. (1996). *Human evolution, language and mind.* Cambridge University Press.

Norris, R. P., Norris, C., Hamacher, D. W., & Abrahams, R. (2013). Wurdi Youang: An Australian Aboriginal stone arrangement with possible solar indications. *Rock Art Research, 30*, 55–65.

Norris, R. P., & Owens, K. (2015, December 23). Keeping track of time. *Australian Academy of Science.*

Nostradamus. (1555). *Les prophéties.* Macé Bonhomme.

Nuclear Energy Agency. (2019). *Preservation of records, knowledge and memory across generations.*

Nunn, P. D., Lancini, L., Franks, L., Compatangelo-Soussignan, R., & McCallum, A. (2019). Maar stories: How oral traditions aid understanding of maar volcanism and associated phenomena during preliterate times. *Annals of the American Association of Geographers, 109*, 1618–1631.

Nyhout, A., & Ganea, P. A. (2019). The development of the counterfactual imagination. *Child Development Perspectives, 13*, 254–259.

O'Callaghan, C., Shine, J. M., Hodges, J. R., Andrews-Hanna, J. R., & Irish, M. (2019). Hippocampal atrophy and intrinsic brain network dysfunction relate to alterations in mind wandering in neurodegeneration. *Proceedings of the National Academy of Sciences, 116*, 3316–3321.

Oeggl, K., Kofler, W., Schmidl, A., Dickson, J. H., Egarter-Vigl, E., & Gaber, O. (2007). The reconstruction of the last itinerary of "Ötzi," the Neolithic iceman, by pollen analyses from sequentially sampled gut extracts. *Quaternary Science Reviews, 26*, 853–861.

Oettingen, G., & Mayer, D. (2002). The motivating function of thinking about the future: Expectations versus fantasies. *Journal of Personality and Social Psychology, 83*, 1198–1212.

Oettingen, G., & Reininger, K. M. (2016). The power of prospection: Mental contrasting and behavior change. *Social and Personality Psychology Compass, 10*, 591–604.

O'Keefe, J., & Dostrovsky, J. (1971). The hippocampus as a spatial map: Preliminary evidence from unit activity in the freely-moving rat. *Brain Research, 34*, 171–175.

O'Keefe, J., & Nadel, L. (1978). *The hippocampus as a cognitive map.* Clarendon Press.

Okuda, J., Fujii, T., Ohtake, H., Tsukiura, T., Tanji, K., Suzuki, K., . . . Yamadori, A. (2003). Thinking of the future and past: The roles of the frontal pole and the medial temporal lobes. *Neuroimage, 19*, 1369–1380.

Oliver, M. (1993, November 24). Larry Walters; soared to fame on lawn chair. *LA Times.*

Oostenbroek, J., Redshaw, J., Davis, J., Kennedy-Costantini, S., Nielsen, M., Slaughter, V., & Suddendorf, T. (2019). Re-evaluating the neonatal imitation hypothesis. *Developmental Science, 22*, e12720.

Oostenbroek, J., Suddendorf, T., Nielsen, M., Redshaw, J., Kennedy-Costantini, S., Davis, J., . . . Slaughter, V. (2016). Comprehensive longitudinal study challenges the existence of neonatal imitation in humans. *Current Biology, 26*, 1334–1338.

Ord, T. (2020). *The precipice: Existential risk and the future of humanity.* Bloomsbury.

Ortega Martínez, A. I., Vallverdú Poch, J., Cáceres, I., Benito-Calvo, A., Parés, J. M., Pérez-Martínez, R., . . . Carbonell, E. (2016, September). *Galería Complex site: The sequence of Acheulean site of Atapuerca (Burgos, Spain).* Paper presented at the European Society for the Study of Human Evolution, Madrid, Spain.

Orwig, J. (2015, January 21). The incredible discovery of the oldest depiction of the universe was almost lost to the black market. *Business Insider.*

Osiurak, F., Lasserre, S., Arbanti, J., Brogniart, J., Bluet, A., Navarro, J., & Reynaud, E. (2021). Technical reasoning is important for cumulative technological culture. *Nature Human Behaviour*, 1–9.

Osiurak, F., & Reynaud, E. (2020). The elephant in the room: What matters cognitively in cumulative technological culture. *Behavioral and Brain Sciences, 43*, 1–57.

O'Sullivan, N. J., Teasdale, M. D., Mattiangeli, V., Maixner, F., Pinhasi, R., Bradley, D. G., & Zink, A. (2016). A whole mitochondria analysis of the Tyrolean iceman's leather provides insights into the animal sources of Copper Age clothing. *Scientific Reports, 6*, 1–9.

Osvath, M. (2009). Spontaneous planning for future stone throwing by a male chimpanzee. *Current Biology, 19*, R190–R191.

Osvath, M., & Gardenfors, P. (2005). Oldowan culture and the evolution of anticipatory cognition. *Lund University Cognitive Studies, 122*, 1–45.

Osvath, M., & Karvonen, E. (2012). Spontaneous innovation for future deception in a male chimpanzee. *PLOS ONE, 7*, e36782.

Owens, R. E. (2008). *Language development: An introduction.* Pearson Education.

Palombo, D. J., Keane, M. M., & Verfaellie, M. (2015). The medial temporal lobes are critical for reward-based decision making under conditions that promote episodic future thinking. *Hippocampus, 25*, 345–353.

Parfit, D. (1984). *Reasons and persons.* Oxford University Press.

Pargeter, J., Khreisheh, N., & Stout, D. (2019). Understanding stone tool-making skill acquisition: Experimental methods and evolutionary implications. *Journal of Human Evolution, 133*, 146–166.

Parker, B. (2011). The tide predictions for D-Day. *Physics Today, 64*, 35–40.

Parker, R. A. (1974). Ancient Egyptian astronomy. *Philosophical Transactions of the Royal Society of London. Series A, Mathematical and Physical Sciences, 276*, 51–65.

Pascoe, B. (2014). *Dark emu: Black seeds: Agriculture or accident?* Magabala Books.

Pearson, J. (2019). The human imagination: The cognitive neuroscience of visual mental imagery. *Nature Reviews Neuroscience, 20*, 624–634.

Peintner, U., Pöder, R., & Pümpel, T. (1998). The iceman's fungi. *Mycological Research, 102*, 1153–1162.

Peng, X., Chen, M., Chen, S., Dasgupta, S., Xu, H., Ta, K., . . . Bai, S. (2018). Microplastics contaminate the deepest part of the world's ocean. *Geochemical Perspectives Letters, 9*, 1–5.

Penn, D. C., Holyoak, K. J., & Povinelli, D. J. (2008). Darwin's mistake: Explaining the discontinuity between human and nonhuman minds. *Behavioral and Brain Sciences, 31*, 109–178.

Perner, J. (1991). *Understanding the representational mind.* MIT Press.

Pernicka, E., Adam, J., Borg, G., Brügmann, G., Bunnefeld, J.-H., Kainz, W., . . . Schwarz, R. (2020). Why the Nebra sky disc dates to the early Bronze Age: An overview of the interdisciplinary results. *Archaeologia Austriaca, 104,* 89–122.

Peters, J., & Büchel, C. (2010). Episodic future thinking reduces reward delay discounting through an enhancement of prefrontal-mediotemporal interactions. *Neuron, 66,* 138–148.

Pfeiffer, B. E., & Foster, D. J. (2013). Hippocampal place-cell sequences depict future paths to remembered goals. *Nature, 497,* 74–79.

Pham, L. B., & Taylor, S. E. (1999). From thought to action: Effects of process-versus outcome-based mental simulations on performance. *Personality and Social Psychology Bulletin, 25,* 250–260.

Piaget, J., & Inhelder, B. (1958). *The growth of logical thinking from childhood to adolescence.* Basic Books.

Pinker, S. (2006). Deep commonalities between life and mind. In A. Grafen & M. Ridley (Eds.), *Richard Dawkins: How a scientist changed the way we think: Reflections by scientists, writers, and philosophers.* Oxford University Press.

Pinker, S. (2010). The cognitive niche: Coevolution of intelligence, sociality, and language. *Proceedings of the National Academy of Sciences, 107,* 8993–8999.

Pinker, S. (2011). *The better angels of our nature: Why violence has declined.* Viking.

Pinker, S. (2018). *Enlightenment now: The case for reason, science, humanism, and progress.* Viking.

Plato. (1952). *Phaedrus* (R. Hackforth, Trans.). Cambridge University Press. (Original work published ca. 360 BC)

Pliny. (1855). *The natural history* (J. Bostock, Trans.). Taylor and Francis. (Original work published 77 AD)

Poirier, L. (2007). Teaching mathematics and the Inuit community. *Canadian Journal of Science, Mathematics and Technology Education, 7,* 53–67.

Polden, J. (2015, November 12). Athlete known as "Monkey Man" runs 100m on all fours in just 15 seconds to break a world record. *Daily Mail.*

Poldrack, R. A. (2018). *The new mind readers: What neuroimaging can and cannot reveal about our thoughts.* Princeton University Press.

Pomeroy, E., Bennett, P., Hunt, C. O., Reynolds, T., Farr, L., Frouin, M., . . . Barker, G. (2020). New Neanderthal remains associated with the "flower burial" at Shanidar Cave. *Antiquity, 94,* 11–26.

Popper, K. (1978). Natural selection and the emergence of mind. *Dialectica, 32,* 339–355.

Possingham, H. (2016). How to make biological conservation a success. *Rundgespräche der Kommission für Ökologie, 44,* 137–142.

Possingham, H., Ball, I., & Andelman, S. (2000). Mathematical methods for identifying representative reserve networks. In S. Ferson & M. Burgman (Eds.), *Quantitative methods for conservation biology.* Springer.

Potts, R., Behrensmeyer, A. K., & Ditchfield, P. (1999). Paleolandscape variation and Early Pleistocene hominid activities: Members 1 and 7, Olorgesailie Formation, Kenya. *Journal of Human Evolution, 37,* 747–788.

Povinelli, D., & Vonk, J. (2003). Chimpanzee minds: Suspiciously human? *Trends in Cognitive Sciences, 7,* 157–160.

Price, M. (2019, December 4). Early humans domesticated themselves, new genetic evidence suggests. *Science*.

Puglise, N. (2016, October 31). Houdini fans hold annual seance: "If anyone could escape the beyond, it's him." *The Guardian*.

Quiroga, R. Q. (2019). Plugging in to human memory: Advantages, challenges, and insights from human single-neuron recordings. *Cell, 179*, 1015–1032.

Quiroga, R. Q. (2021). How are memories stored in the human hippocampus? *Trends in Cognitive Sciences, 25*, 425–426.

Radovčić, D., Sršen, A. O., Radovčić, J., & Frayer, D. W. (2015). Evidence for Neandertal jewelry: Modified white-tailed eagle claws at Krapina. *PLOS ONE, 10*, e0119802.

Rafetseder, E., Cristi-Vargas, R., & Perner, J. (2010). Counterfactual reasoning: Developing a sense of "nearest possible world." *Child Development, 81*, 376–389.

Rafetseder, E., & Perner, J. (2014). Counterfactual reasoning: Sharpening conceptual distinctions in developmental studies. *Child Development Perspectives, 8*, 54–58.

Rajecki, D. W. (1974). Effects of prenatal exposure to auditory or visual stimulation on postnatal distress vocalizations in chicks. *Behavioral Biology, 11*, 525–536.

Ramanan, S., Piguet, O., & Irish, M. (2018). Rethinking the role of the angular gyrus in remembering the past and imagining the future: The contextual integration model. *Neuroscientist, 24*, 342–352.

Rao, R. P., & Ballard, D. H. (1999). Predictive coding in the visual cortex: A functional interpretation of some extra-classical receptive-field effects. *Nature Neuroscience, 2*, 79–87.

Read, D. W. (2008). Working memory: A cognitive limit to non-human primate recursive thinking prior to hominid evolution. *Evolutionary Psychology, 6*, 676–714.

Read, V. M. S. J., & Hughes, R. N. (1987). Feeding behaviour and prey choice in *Macroperipatus torquatus* (Onychophora). *Proceedings of the Royal Society B: Biological Sciences, 230*, 483–506.

Reddan, M. C., Wager, T. D., & Schiller, D. (2018). Attenuating neural threat expression with imagination. *Neuron, 100*, 994–1005.

Redish, A. D. (2016). Vicarious trial and error. *Nature Reviews Neuroscience, 17*, 147–159.

Redshaw, J. (2014). Does metarepresentation make human mental time travel unique? *Wiley Interdisciplinary Reviews: Cognitive Science, 5*, 519–531.

Redshaw, J. (2019). Re-analysis of data reveals no evidence for neonatal imitation in rhesus macaques. *Biology Letters, 15*, 20190342.

Redshaw, J., & Bulley, A. (2018). Future-thinking in animals: Capacities and limits. In G. Oettingen, A. T. Servincer, & P. M. Gollwitzer (Eds.), *The psychology of thinking about the future*. Guilford Press.

Redshaw, J., Bulley, A., & Suddendorf, T. (2019). Thinking about thinking about time. *Behavioral and Brain Sciences, 42*, e273.

Redshaw, J., Leamy, T., Pincus, P., & Suddendorf, T. (2018). Young children's capacity to imagine and prepare for certain and uncertain future outcomes. *PLOS ONE, 13*, e0202606.

Redshaw, J., Nielsen, M., Slaughter, V., Kennedy-Costantini, S., Oostenbroek, J., Crimston, J., & Suddendorf, T. (2020). Individual differences in neonatal "imitation" fail to predict early social cognitive behaviour. *Developmental Science, 23*, e12892.

Redshaw, J., & Suddendorf, T. (2013). Foresight beyond the very next event: Four-year-olds can link past and deferred future episodes. *Frontiers in Psychology, 4*, 404.

Redshaw, J., & Suddendorf, T. (2016). Children's and apes' preparatory responses to two mutually exclusive possibilities. *Current Biology, 26*, 1758–1762.

Redshaw, J., & Suddendorf, T. (2018). Misconceptions about adaptive function. *Behavioral and Brain Sciences, 41*, 38–39.

Redshaw, J., & Suddendorf, T. (2020). Temporal junctures in the mind. *Trends in Cognitive Sciences, 24*, 52–64.

Redshaw, J., Suddendorf, T., Neldner, K., Wilks, M., Tomaselli, K., Mushin, I., & Nielsen, M. (2019). Young children from three diverse cultures spontaneously and consistently prepare for alternative future possibilities. *Child Development, 90*, 51–61.

Redshaw, J., Taylor, A. H., & Suddendorf, T. (2017). Flexible planning in ravens? *Trends in Cognitive Sciences, 21*, 821–822.

Reich, D., Patterson, N., Kircher, M., Delfin, F., Nandineni, M. R., Pugach, I., . . . Stoneking, M. (2011). Denisova admixture and the first modern human dispersals into Southeast Asia and Oceania. *American Journal of Human Genetics, 89*, 516–528.

Rescorla, R. A., & Wagner, A. R. (1972). A theory of Pavlovian conditioning: Variations in the effectiveness of reinforcement and nonreinforcement. In A. H. Black & W. F. Prokasy (Eds.), *Classical conditioning II: Current research and theory*. Appleton-Century-Crofts.

Ritchie, H. (2019, November 11). Half of the world's habitable land is used for agriculture. *Our World in Data*.

Ritchie, H., & Roser, M. (2018, June). Ozone Layer. *Our World in Data*.

Roberts, P., & Stewart, B. A. (2018). Defining the "generalist specialist" niche for Pleistocene *Homo sapiens*. *Nature Human Behaviour, 2*, 542–550.

Robinson, E. J., Rowley, M. G., Beck, S. R., Carroll, D. J., & Apperly, I. A. (2006). Children's sensitivity to their own relative ignorance: Handling of possibilities under epistemic and physical uncertainty. *Child Development, 77*, 1642–1655.

Robson, S. L., & Wood, B. (2008). Hominin life history: Reconstruction and evolution. *Journal of Anatomy, 212*, 394–425.

Roesch, M. R., Esber, G. R., Li, J., Daw, N. D., & Schoenbaum, G. (2012). Surprise! Neural correlates of Pearce–Hall and Rescorla–Wagner coexist within the brain. *European Journal of Neuroscience, 35*, 1190–1200.

Rogers, N., Killcross, S., & Curnoe, D. (2016). Hunting for evidence of cognitive planning: Archaeological signatures versus psychological realities. *Journal of Archaeological Science: Reports, 5*, 225–239.

Roksandic, M., Radović, P., Wu, X. J., & Bae, C. J. (2022). Resolving the "muddle in the middle": The case for *Homo bodoensis* sp. nov. *Evolutionary Anthropology, 31*, 20–29.

Rösch, S. A., Stramaccia, D. F., & Benoit, R. G. (in press). Promoting farsighted decisions via episodic future thinking: A meta-analysis. *Journal of Experimental Psychology: General*.

Rosenbaum, R. S., Kohler, S., Schacter, D. L., Moscovitch, M., Westmacott, R., Black, S. E., . . . Tulving, E. (2005). The case of KC: Contributions of a memory-impaired person to memory theory. *Neuropsychologia, 43*, 989–1021.

Rosling, H., Rosling, O., & Rosling Rönnlund, A. (2018). *Factfulness*. Flatiron Books.

Roth, G., & Dicke, U. (2005). Evolution of the brain and intelligence. *Trends in Cognitive Sciences, 9*, 250–257.

Rousseau, J.-J. (1913). *The social contract* (G. D. H. Cole, Trans.). E. P. Dutton & Co. (Original work published 1762)

Ruby, F. J., Smallwood, J., Sackur, J., & Singer, T. (2013). Is self-generated thought a means of social problem solving? *Frontiers in Psychology, 4*, 962.

Ruggles, C. (1997). Astronomy and Stonehenge. *Proceedings of the British Academy, 92*, 203–230.

Ruginski, I. T., Creem-Regehr, S. H., Stefanucci, J. K., & Cashdan, E. (2019). GPS use negatively affects environmental learning through spatial transformation abilities. *Journal of Environmental Psychology, 64*, 12–20.

Rust, S. (2019, November 10). How the U.S. betrayed the Marshall Islands, kindling the next nuclear disaster. *Los Angeles Times*.

Ruxton, G. D., & Hansell, M. H. (2011). Fishing with a bait or lure: A brief review of the cognitive issues. *Ethology, 117*, 1–9.

Safire, W. (1969, July 18). Memo in the event of a moon disaster. US National Archives.

Sagan, C. (1994). *Pale blue dot: A vision of the human future in space*. Random House.

Sahle, Y., Hutchings, W. K., Braun, D. R., Sealy, J. C., Morgan, L. E., Negash, A., & Atnafu, B. (2013). Earliest stone-tipped projectiles from the Ethiopian Rift date to >279,000 years ago. *PLOS ONE, 8*, e78092.

Sala, N., Arsuaga, J. L., Pantoja-Pérez, A., Pablos, A., Martínez, I., Quam, R. M., . . . Carbonell, E. (2015). Lethal interpersonal violence in the Middle Pleistocene. *PLOS ONE, 10*, e0126589.

Salk, J. (1992). Are we being good ancestors? *World Affairs: The Journal of International Issues, 1*, 16–18.

Samuelson, A. (2019, June 20). *What is an atomic clock?* NASA.

Sánchez, O., Vargas, J. A., & López-Forment, W. (1999). Observations of bats during a total solar eclipse in Mexico. *Southwestern Naturalist, 44*, 112–115.

Sandom, C., Faurby, S., Sandel, B., & Svenning, J.-C. (2014). Global late Quaternary megafauna extinctions linked to humans, not climate change. *Proceedings of the Royal Society B: Biological Sciences, 281*, 20133254.

Santos, L. R., & Rosati, A. G. (2015). The evolutionary roots of human decision making. *Annual Review of Psychology, 66*, 321–347.

Sapolsky, R. M. (2004). *Why zebras don't get ulcers: The acclaimed guide to stress, stress-related diseases, and coping*. Holt Paperbacks.

Saturno, W. A., Stuart, D., & Beltrán, B. (2006). Early Maya writing at San Bartolo, Guatemala. *Science, 311*, 1281–1283.

Schacter, D. L. (1999). The seven sins of memory: Insights from psychology and cognitive neuroscience. *American Psychologist, 54*, 182–203.

Schacter, D. L., Addis, D. R., & Buckner, R. L. (2007). Remembering the past to imagine the future: The prospective brain. *Nature Reviews Neuroscience, 8*, 657–661.

Schacter, D. L., Addis, D. R., Hassabis, D., Martin, V., Spreng, R., & Szpunar, K. (2012). The future of memory: Remembering, imagining, and the brain. *Neuron, 76*, 677–694.

Schacter, D. L., Devitt, A. L., & Addis, D. R. (2018). Episodic future thinking and cognitive aging. In B. Knight (Ed.), *Oxford research encyclopedia of psychology*. Oxford University Press.

Schmandt-Besserat, D. (1981). Decipherment of the earliest tablets. *Science, 211*, 283–285.

Schmidt, P., Blessing, M., Rageot, M., Iovita, R., Pfleging, J., Nickel, K. G., . . . Tennie, C. (2019). Birch tar production does not prove Neanderthal behavioral complexity. *Proceedings of the National Academy of Sciences, 116*, 17707–17711.

Schopenhauer, A. (1913). *Studies in pessimism* (T. B. Saunders, Trans.). G. Allen. (Original work published 1851)

Schultz, W. (1998). Predictive reward signal of dopamine neurons. *Journal of Neurophysiology, 80,* 1–27.

Schuppli, C., & van Schaik, C. P. (2019). Animal cultures: How we've only seen the tip of the iceberg. *Evolutionary Human Sciences, 1,* e2.

Science. (2007). Breakthrough of the year: The runners-up. *Science, 318,* 1844–1849.

Scott, J. C. (2017). *Against the grain.* Yale University Press.

Secretariat of the Convention on Biological Diversity. (2020). *Global Biodiversity Outlook 5.*

Seiradakis, J. H., & Edmunds, M. G. (2018). Our current knowledge of the Antikythera mechanism. *Nature Astronomy, 2,* 35–42.

Seli, P., Carriere, J. S., Wammes, J. D., Risko, E. F., Schacter, D. L., & Smilek, D. (2018). On the clock: Evidence for the rapid and strategic modulation of mind wandering. *Psychological Science, 29,* 1247–1256.

Seli, P., Risko, E.F., Smilek, D., & Schacter, D. L. (2016). Mind-wandering with and without intention. *Trends in Cognitive Sciences, 20,* 605–617.

Seligman, M. E. P., Railton, P., Baumeister, R. F., & Sripada, C. (2016). *Homo prospectus.* Oxford University Press.

Seneca. (2017). *Dialogues and essays* (J. Davie, Trans.). Oxford University Press. (Original work published ca. 64 AD)

Seneca. (1969). *Letters from a stoic* (R. Campbell, Trans.). Penguin Books. (Original work published ca. 65 AD)

Seth, A. K. (2019). Our inner universes. *Scientific American, 321,* 40–47.

Shahack-Gross, R., Berna, F., Karkanas, P., Lemorini, C., Gopher, A., et al. (2014). Evidence for the repeated use of a central hearth of Middle Pleistocene (300 ky ago) Qesem Cave, Israel. *Journal of Archaeological Science, 44,* 12–21.

Sharot, T. (2011). The optimism bias. *Current Biology, 21,* R941–R945.

Sharot, T., & Garrett, N. (2016). Forming beliefs: Why valence matters. *Trends in Cognitive Sciences, 20,* 25–33.

Sharot, T., Korn, C. W., & Dolan, R. J. (2011). How unrealistic optimism is maintained in the face of reality. *Nature Neuroscience, 14,* 1475–1479.

Sharpe, L. L. (2019). Fun, fur, and future fitness: The evolution of play in mammals. In P. K. Smith & J. L. Roopnarine (Eds.), *The Cambridge handbook of play: Developmental and disciplinary perspectives.* Cambridge University Press.

Shaw, M. J. (2011). *Time and the French Revolution: The republican calendar, 1789–Year xiv.* Royal Historical Society.

Shipton, C. (2020). The unity of Acheulean culture. In H. S. Groucutt (Ed.), *Culture history and convergent evolution: Can we detect populations in prehistory?* Springer International Publishing.

Shipton, C., & Nielsen, M. (2015). Before cumulative culture: The evolutionary origins of overimitation and shared intentionality. *Human Nature, 26,* 331–345.

Silberberg, A., & Kearns, D. (2009). Memory for the order of briefly presented numerals in humans as a function of practice. *Animal Cognition, 12,* 405–407.

Skinner, B. F. (1948). "Superstition" in the pigeon. *Journal of Experimental Psychology, 38,* 168–172.

Skinner, B. F. (1953). *Science and human behaviour*. Macmillan.

Skinner, B. F. (1969). *Walden two*. Hackett Publishing.

Slaughter, V., & Boh, W. (2001). Decalage in infants' search for mothers versus toys demonstrated with a delayed response task. *Infancy, 2*, 405–413.

Slaughter, V., & Griffiths, M. (2007). Death understanding and fear of death in young children. *Clinical Child Psychology and Psychiatry, 12*, 525–535.

Smaers, J. B., Gómez-Robles, A., Parks, A. N., & Sherwood, C. C. (2017). Exceptional evolutionary expansion of prefrontal cortex in great apes and humans. *Current Biology, 27*, 714–720.

Smallwood, J., Bernhardt, B. C., Leech, R., Bzdok, D., Jefferies, E., & Margulies, D. S. (2021). The default mode network in cognition: A topographical perspective. *Nature Reviews Neuroscience, 22*, 503–513.

Smallwood, J., & Schooler, J. W. (2015). The science of mind wandering: Empirically navigating the stream of consciousness. *Annual Review of Psychology, 66*, 487–518.

Smith, A. (2011). The Chinese sexagenary cycle and the ritual origins of the calendar. In J. Steele (Ed.), *Calendars and years II: Astronomy and time in the ancient and medieval world*. Oxbox Books.

Smith, F. A., Smith, R. E. E., Lyons, S. K., & Payne, J. L. (2018). Body size downgrading of mammals over the late Quaternary. *Science, 360*, 310–313.

Smolker, R., Richards, A., Connor, R., Mann, J., & Berggren, P. (1997). Sponge carrying by dolphins (Delphinidae, *Tursiops* sp.): A foraging specialization involving tool use? *Ethology, 103*, 454–465.

Solda, A., Ke, C., Page, L., & von Hippel, W. (2020). Strategically delusional. *Experimental Economics, 23*, 604–631.

Southern, L. M., Deschner, T., & Pika, S. (2021). Lethal coalitionary attacks of chimpanzees (*Pan troglodytes troglodytes*) on gorillas (*Gorilla gorilla gorilla*) in the wild. *Scientific Reports, 11*, 14673.

Soutschek, A., Ugazio, G., Crockett, M. J., Ruff, C. C., Kalenscher, T., & Tobler, P. N. (2017). Binding oneself to the mast: Stimulating frontopolar cortex enhances precommitment. *Social Cognitive and Affective Neuroscience, 12*, 635–642.

Spelke, E. S., & Kinzler, K. D. (2007). Core knowledge. *Developmental Science, 10*, 89–96.

Spencer, H. (1862). *First principles* (Fourth ed.). D. Appleton and Company.

Spinoza, B. (2018). *The ethics* (R. H. M. Elwes, Trans.). Dover Publications, Inc. (Original work published 1677)

Stedman, H. H., Kozyak, B. W., Nelson, A., Thesier, D. M., Su, L. T., Low, D. W., . . . Mitchell, M. A. (2004). Myosin gene mutation correlates with anatomical changes in the human lineage. *Nature, 428*, 415–418.

Steele, J. (2017). *Rising time schemes in Babylonian astronomy*. Springer.

Steele, J. (2021). The continued relevance of MUL.APIN in late Babylonian astronomy. *Journal of Ancient Near Eastern History, 8*, 259–277.

Steffen, W., Richardson, K., Rockström, J., Cornell, S. E., Fetzer, I., Bennett, E. M., . . . Sörlin, S. (2015). Planetary boundaries: Guiding human development on a changing planet. *Science, 347*, 1259855.

Sterelny, K. (2007). Social intelligence, human intelligence and niche construction. *Philosophical Transactions of the Royal Society B: Biological Sciences, 362*, 719–730.

Sterelny, K. (2012). *The evolved apprentice: How evolution made humans unique.* MIT Press.

Stout, D., Rogers, M. J., Jaeggi, A. V., & Semaw, S. (2019). Archaeology and the origins of human cumulative culture: A case study from the earliest Oldowan at Gona, Ethiopia. *Current Anthropology, 60,* 309–340.

Strikwerda-Brown, C., Grilli, M. D., Andrews-Hanna, J., & Irish, M. (2019). "All is not lost"—Rethinking the nature of memory and the self in dementia. *Ageing Research Reviews, 54,* 100932.

Stringer, C., Grün, R., Schwarcz, H., & Goldberg, P. (1989). ESR dates for the hominid burial site of Es Skhul in Israel. *Nature, 338,* 756–758.

Suddendorf, T. (1994). *Discovery of the fourth dimension: Mental time travel and human evolution.* (Master of Social Sciences in Psychology). University of Waikato.

Suddendorf, T. (1999). The rise of the metamind. In M. C. Corballis & S. E. G. Lea (Eds.), *The descent of mind: Psychological perspectives on hominid evolution.*

Suddendorf, T. (2006). Foresight and evolution of the human mind. *Science, 312,* 1006–1007.

Suddendorf, T. (2010). Linking yesterday and tomorrow: Preschoolers' ability to report temporally displaced events. *British Journal of Developmental Psychology, 28,* 491–498.

Suddendorf, T. (2011). Evolution, lies and foresight biases. *Behavioral and Brain Sciences, 34,* 38–39.

Suddendorf, T. (2013a). *The gap: The science of what separates us from other animals.* Basic Books.

Suddendorf, T. (2013b). Mental time travel: Continuities and discontinuities. *Trends in Cognitive Sciences, 17,* 151–152.

Suddendorf, T. (2017). The emergence of episodic foresight and its consequences. *Child Development Perspectives, 11,* 191–195.

Suddendorf, T., Brinums, M., & Imuta, K. (2016). Shaping one's future self: The development of deliberate practice. In S. B. Klein, K. Micheaelian, & K. Szpunar (Eds.), *Seeing the future: Theoretical perspectives on future-oriented mental time travel.* Oxford University Press.

Suddendorf, T., Bulley, A., & Miloyan, B. (2018). Prospection and natural selection. *Current Opinion in Behavioral Sciences, 24,* 26–31.

Suddendorf, T., & Busby, J. (2003). Mental time travel in animals? *Trends in Cognitive Sciences, 7,* 391–396.

Suddendorf, T., & Busby, J. (2005). Making decisions with the future in mind: Developmental and comparative identification of mental time travel. *Learning and Motivation, 36,* 110–125.

Suddendorf, T., & Butler, D. L. (2013). The nature of visual self-recognition. *Trends in Cognitive Sciences, 17,* 121–127.

Suddendorf, T., & Corballis, M. C. (1997). Mental time travel and the evolution of the human mind. *Genetic Social and General Psychology Monographs, 123,* 133–167.

Suddendorf, T., & Corballis, M. C. (2007). The evolution of foresight: What is mental time travel and is it unique to humans? *Behavioral and Brain Sciences, 30,* 299–313.

Suddendorf, T., & Corballis, M. C. (2010). Behavioural evidence for mental time travel in nonhuman animals. *Behavioural Brain Research, 215,* 292–298.

Suddendorf, T., Crimston, J., & Redshaw, J. (2017). Preparatory responses to socially determined, mutually exclusive possibilities in chimpanzees and children. *Biology Letters, 13,* 20170170.

Suddendorf, T., Kirkland, K., Bulley, A., Redshaw, J., & Langley, M. C. (2020). It's in the bag: Mobile containers in human evolution and child development. *Evolutionary Human Sciences, 2,* e48.

Suddendorf, T., & Moore, C. (2011). Introduction to the special issue: The development of episodic foresight. *Cognitive Development, 26,* 295–298.

Suddendorf, T., Nielsen, M., & Von Gehlen, R. (2011). Children's capacity to remember a novel problem and to secure its future solution. *Developmental Science, 14,* 26–33.

Suddendorf, T., & Redshaw, J. (2013). The development of mental scenario building and episodic foresight. *Annals of the New York Academy of Sciences,* 1296, 135–153.

Suddendorf, T., Watson, K., Bogaart, M., & Redshaw, J. (2020). Preparation for certain and uncertain future outcomes in young children and three species of monkey. *Developmental Psychobiology, 62,* 191–201.

Suddendorf, T., & Whiten, A. (2001). Mental evolution and development: Evidence for secondary representation in children, great apes and other animals. *Psychological Bulletin, 127,* 629–650.

Suetonius. (1957). *The twelve Caesars* (R. Graves, Trans.). Penguin Books. (Original work published 121 AD)

Sun Tzu. (1910). *The art of war* (L. Giles, Trans.). Luzac. (Original work ca. fifth century BC)

Sutton, J. (2010). Exograms and interdisciplinarity: History, the extended mind, and the civilizing process. In R. Menary (Ed.), *The extended mind.* MIT Press.

Sutton, P., & Walshe, K. (2021). *Farmers or hunter-gatherers? The Dark Emu debate.* Melbourne University Publishing.

Swerdlow, N. M., & Neugebauer, O. (1984). *Mathematical astronomy in Copernicus's De Revolutionibus.* Springer-Verlag.

Szpunar, K. K., & Schacter, D. L. (2013). Get real: Effects of repeated simulation and emotion on the perceived plausibility of future experiences. *Journal of Experimental Psychology: General, 142,* 323–327.

Tagkopoulos, I., Liu, Y. C., & Tavazoie, S. (2008). Predictive behavior within microbial genetic networks. *Science, 320,* 1313–1317.

Tang, M. F., Smout, C. A., Arabzadeh, E., & Mattingley, J. B. (2018). Prediction error and repetition suppression have distinct effects on neural representations of visual information. *eLife, 7,* e33123.

Taungurung Land and Waters Council. (2021). *Waang the trickster.*

Taylor, A. H., Elliffe, D., Hunt, G. R., & Gray, R. D. (2010). Complex cognition and behavioural innovation in New Caledonian crows. *Proceedings of the Royal Society B: Biological Sciences, 277,* 2637–2643.

Taylor, A. H., Medina, F. S., Holzhaider, J. C., Hearne, L. J., Hunt, G. R., & Gray, R. D. (2010). An investigation into the cognition behind spontaneous string pulling in New Caledonian crows. *PLOS ONE, 5,* e9345.

Taylor, A. H., Miller, R., & Gray, R. D. (2012). New Caledonian crows reason about hidden causal agents. *Proceedings of the National Academy of Sciences, 109,* 16389–16391.

Tecwyn, E. C., Thorpe, S. K. S., & Chappell, J. (2013). A novel test of planning ability: Great apes can plan step-by-step but not in advance of action. *Behavioural Processes, 100,* 174–184.

Teeple, J. E. (1926). Maya inscriptions: The Venus calendar and another correlation. *American Anthropologist, 28,* 402–408.

Tennie, C., Call, J., & Tomasello, M. (2009). Ratcheting up the ratchet: On the evolution of cumulative culture. *Philosophical Transactions of the Royal Society B: Biological Sciences, 364,* 2405–2415.

Terrett, G., Lyons, A., Henry, J. D., Ryrie, C., Suddendorf, T., & Rendell, P. G. (2017). Acting with the future in mind is impaired in long-term opiate users. *Psychopharmacology, 234,* 99–108.

Tetlock, P. E., & Gardner, D. (2016). *Superforecasting: The art and science of prediction.* Random House.

Thakral, P. P., Madore, K. P., & Schacter, D. L. (2017). A role for the left angular gyrus in episodic simulation and memory. *Journal of Neuroscience, 37,* 8142–8149.

Thieme, H. (1997). Lower Palaeolithic hunting spears from Germany. *Nature, 385,* 807–810.

Thomas, N. W. (1924). The week in West Africa. *Journal of the Royal Anthropological Institute of Great Britain and Ireland, 54,* 183–209.

Thompson, J. (1972). *A Commentary on the Dresden Codex: A Maya hieroglyphic book.* American Philosophical Society.

Thorndike, E. L. (1898). Animal intelligence: An experimental study of the associative process in animals. *Psychological Review and Monograph, 2,* 551–553.

Thornton, A., & McAuliffe, K. (2006). Teaching in wild meerkats. *Science, 313,* 227–229.

Thorpe, I. J. N. (2003). Anthropology, archaeology, and the origin of warfare. *World Archaeology, 35,* 145–165.

Tinbergen, N. (1963). On aims and methods of ethology. *Zeitschrift fuer Tierpsychologie, 20,* 410–433.

Tobin, H., & Logue, A. W. (1994). Self-control across species (*Columba livia, Homo sapiens,* and *Rattus norvegicus*). *Journal of Comparative Psychology, 108,* 126–133.

Tolman, E. C. (1939). Prediction of vicarious trial and error by means of the schematic sowbug. *Psychological Review, 46,* 318–336.

Tolman, E. C. (1948). Cognitive maps in rats and men. *Psychological Review, 55,* 189–208.

Tomasello, M. (2019). *Becoming human.* Harvard University Press.

Tomasello, M., Call, J., & Hare, B. (2003). Chimpanzees understand psychological states—the question is which ones and to what extent. *Trends in Cognitive Sciences, 7,* 153–156.

Tooby, J., & DeVore, I. (1987). The reconstruction of hominid behavioral evolution through strategic modelling. In W. Kinzey (Ed.), *The evolution of human behavior: Primate models.* State University of New York Press.

Toth, N., & Schick, K. (2018). An overview of the cognitive implications of the Oldowan Industrial Complex. *Azania: Archaeological Research in Africa, 53,* 3–39.

Trope, Y., & Liberman, N. (2010). Construal-level theory of psychological distance. *Psychological Review, 117,* 440–463.

Tuckerman, J. (2019, November 7). Mexican mammoth trap provides first evidence of prehistoric hunting pits. *The Guardian.*

Tulving, E. (1985). Memory and consciousness. *Canadian Psychology, 26*, 1–12.

Tulving, E. (2005). Episodic memory and autonoesis: Uniquely human? In H. S. Terrace & J. Metcalfe (Eds.), *The missing link in cognition: Evolution of self-knowing consciousness.* Oxford University Press.

Tversky, A., & Kahneman, D. (1973). Availability heuristic for judging frequency and probability. *Cognitive Psychology, 5*, 207–232.

Tyerman, D., & Bennet, G. (1831). *Journal of voyages and travels.* Frederick Westley and A. H. Davis.

Uetz, G. W., Hieber, C. S., Jakob, E. M., Wilcox, R. S., Kroeger, D., McCrate, A., & Mostrom, A. M. (1994). Behavior of colonial orb-weaving spiders during a solar eclipse. *Ethology, 96*, 24–32.

Ulber, J., Hamann, K., & Tomasello, M. (2017). Young children, but not chimpanzees, are averse to disadvantageous and advantageous inequities. *Journal of Experimental Child Psychology, 155*, 48–66.

Umbach, G., Kantak, P., Jacobs, J., Kahana, M., Pfeiffer, B. E., Sperling, M., & Lega, B. (2020). Time cells in the human hippocampus and entorhinal cortex support episodic memory. *Proceedings of the National Academy of Sciences, 117*, 28463–28474.

Unknown author. (ca. thirty-first century BC). Beer production at the inanna temple in Uruk. Schøyen Collection, Oslo, Norway.

Urton, G. (2003). *Signs of the Inka khipu: Binary coding in the Andean knotted-string records.* University of Texas Press.

Utrilla, P., Mazo, C., Domingo, R., & Bea, M. (2021). Maps in prehistoric art. In I. Davidson & A. Nowell (Eds.), *Making scenes: Global perspectives on scenes in rock art.* Berghahn Books.

Vail, G. (2006). The Maya codices. *Annual Review of Anthropology, 35*, 497–519.

Vale, G. L., Flynn, E. G., & Kendal, R. L. (2012). Cumulative culture and future thinking: Is mental time travel a prerequisite to cumulative cultural evolution? *Learning and Motivation, 43*, 220–230.

van den Bergh, G. D., Kaifu, Y., Kurniawan, I., Kono, R. T., Brumm, A., Setiyabudi, E., . . . Morwood, M. J. (2016). *Homo floresiensis*-like fossils from the early Middle Pleistocene of Flores. *Nature, 534*, 245–248.

van Schaik, C. P., Damerius, L., & Isler, K. (2013). Wild orangutan males plan and communicate their travel direction one day in advance. *PLOS ONE, 8*, e74896.

Velpeau, A. A. L. M. (1839). *Nouveaux éléments de médecine opératoire.* J.-B. Baillière.

Venkataraman, B. (2019). *The optimist's telescope: Thinking ahead in a reckless age.* Riverhead Books.

Villar, R. (2012, April 18). No monkeying around for Japan man, fastest on four legs. *Reuters.*

Vince, G. (2019). *Transcendence: How humans evolved through fire, language, beauty, and time.* Penguin UK.

Visser, I., Smith, T., Bullock, I., Green, G., Carlsson, O. L., & Imberti, S. (2008). Antarctic peninsula killer whales (*Orcinus orca*) hunt seals and a penguin on floating ice. *Marine Mammal Science, 24*, 225–234.

Völter, C. J., Mundry, R., Call, J., & Seed, A. M. (2019). Chimpanzees flexibly update working memory contents and show susceptibility to distraction in the self-ordered search task. *Proceedings of the Royal Society B, 286*, 20190715.

von Hippel, E., De Jong, J. P., & Flowers, S. (2012). Comparing business and household sector innovation in consumer products: Findings from a representative study in the United Kingdom. *Management Science, 58*, 1669–1681.

von Hippel, W. (2018). *The social leap.* HarperCollins.

von Hippel, W., & Suddendorf, T. (2018). Did humans evolve to innovate with a social rather than technical orientation? *New Ideas in Psychology, 51*, 34–39.

Vonk, J. (2020). Twenty years after folk physics for apes: Researchers' understanding of how nonhumans understand the world. *Animal Behavior and Cognition, 7*, 264–269.

Wahba, M. A., & Bridwell, L. G. (1976). Maslow reconsidered: A review of research on the need hierarchy theory. *Organizational Behavior and Human Performance, 15*, 212–240.

Walker, C. M., & Gopnik, A. (2014). Toddlers infer higher-order relational principles in causal learning. *Psychological Science, 25*, 161–169.

Walker, R. S., & Bailey, D. H. (2013). Body counts in lowland South American violence. *Evolution and Human Behavior, 34*, 29–34.

Wallace, A. R. (1870). *Contributions to the theory of natural selection: A series of essays.* Macmillan and Company.

Ward, J. (2014). *Adventures in stationery: A journey through your pencil case.* Profile Books.

Watabe-Uchida, M., Eshel, N., & Uchida, N. (2017). Neural circuitry of reward prediction error. *Annual Review of Neuroscience, 40*, 373–394.

Waters, C. N., Zalasiewicz, J., Summerhayes, C., Barnosky, A. D., Poirier, C., Gałuszka, A., . . . Wolfe, A. P. (2016). The Anthropocene is functionally and stratigraphically distinct from the Holocene. *Science, 351*, aad2622.

Watts, T. W., Duncan, G. J., & Quan, H. (2018). Revisiting the marshmallow test: A conceptual replication investigating links between early delay of gratification and later outcomes. *Psychological Science, 29*, 1159–1177.

Wearing, D. (2005). *Forever today: A memoir of love and amnesia.* Random House.

Weimer, A. A., Sallquist, J., & Bolnick, R. R. (2012). Young children's emotion comprehension and theory of mind understanding. *Early Education and Development, 23*, 280–301.

Weiner, N. (1950). *The human use of human beings.* Houghton Mifflin.

Weir, A. A. S., Chappell, J., & Kacelnik, A. (2002). Shaping of hooks in New Caledonian crows. *Science, 297*, 981.

Weisberg, J. (2018). *Asking for a friend: Three centuries of advice on life, love, money, and other burning questions from a nation obsessed.* Nation Books.

Wellman, H. M., Cross, D., & Watson, J. (2001). Meta-analysis of theory-of-mind development: The truth about false belief. *Child Development, 72*, 655–684.

Wellman, H. M., Fang, F., & Peterson, C. C. (2011). Sequential progressions in a theory-of-mind scale: Longitudinal perspectives. *Child Development, 82*, 780–792.

Wellman, H. M., & Liu, D. (2004). Scaling of theory-of-mind tasks. *Child Development, 75*, 523–541.

Wells, A. (2005). The metacognitive model of GAD: Assessment of meta-worry and relationship with DSM-IV generalized anxiety disorder. *Cognitive Therapy and Research, 29*, 107–121.

Wells, G. (1985). *Language development in the pre-school years.* Cambridge University Press.

Wells, H. G. (1895). *The time machine.* William Heineman.

White, M., & Ashton, N. (2003). Lower Palaeolithic core technology and the origins of the Levallois method in North-Western Europe. *Current Anthropology, 44*, 598–609.

White, R., Bosinski, G., Bourrillon, R., Clottes, J., Conkey, M., Rodriguez, S. C., . . . Delluc, G. (2020). Still no archaeological evidence that Neanderthals created Iberian cave art. *Journal of Human Evolution, 144*, 102640.

Whiten, A. (1999). The evolution of deep social mind in humans. In M. C. Corballis & S. E. G. Lea (Eds.), *The descent of mind: Psychological perspectives on hominid evolution.* Oxford University Press.

Whiten, A. (2021). The burgeoning reach of animal culture. *Science, 372*, eabe6514.

Whiten, A., & Erdal, D. (2012). The human socio-cognitive niche and its evolutionary origins. *Philosophical Transactions of the Royal Society B: Biological Sciences, 367*, 2119–2129.

Whiten, A., Goodall, J., McGrew, W. C., Nishida, T., Reynolds, V., Sugiyama, Y., . . . Boesch, C. (1999). Cultures in chimpanzees. *Nature, 399*, 682–685.

Wierenga, J. (2001, April 12). Kano van pesse kon echt varen. *Nieuwsblad van het Noorden.*

Wierer, U., Arrighi, S., Bertola, S., Kaufmann, G., Baumgarten, B., Pedrotti, A., . . . Pelegrin, J. (2018). The iceman's lithic toolkit: Raw material, technology, typology and use. *PLOS ONE, 13*, e0198292.

Wiessner, P. W. (2014). Embers of society: Firelight talk among the Ju/'hoansi Bushmen. *Proceedings of the National Academy of Sciences, 111*, 14027–14035.

Wilkins, J., Schoville, B. J., Brown, K. S., & Chazan, M. (2012). Evidence for early hafted hunting technology. *Science, 338*, 942–946.

Williams, D. (2018). Predictive minds and small-scale models: Kenneth Craik's contribution to cognitive science. *Philosophical Explorations, 21*, 245–263.

Willms, A. R., Kitanov, P. M., & Langford, W. F. (2017). Huygens' clocks revisited. *Royal Society Open Science, 4*, 170777.

Wimmer, H., & Perner, J. (1983). Beliefs about beliefs: Representation and constraining function of wrong beliefs in young children's understanding of deception. *Cognition, 13*, 103–128.

Wittmann, M. (2016). *Felt time: The psychology of how we perceive time.* MIT Press.

Witze, A. (2021). NASA spacecraft will slam into asteroid in first planetary-defence test. *Nature, 600*, 17–18.

Wrangham, R. (2009). *Catching fire: How cooking made us human.* Basic Books.

Wynne, C. D. L. (2004). Fair refusal by capuchin monkeys. *Nature, 428*, 140.

Ye, J.-y., Qin, X.-j., Cui, J.-f., Ren, Q., Jia, L.-x., Wang, Y., . . . Chan, R. C. (2021). A meta-analysis of mental time travel in individuals with autism spectrum disorders. *Journal of Autism and Developmental Disorders.*

Yellen, J. E., Brooks, A. S., Cornelissen, E., Mehlman, M. J., & Stewart, K. (1995). A Middle Stone Age worked bone industry from Katanda, Upper Semliki Valley, Zaire. *Science, 268*, 553–556.

Young, R. W. (2003). Evolution of the human hand: The role of throwing and clubbing. *Journal of Anatomy, 202*, 165–174.

Zalasiewicz, J., Williams, M., Waters, C. N., Barnosky, A. D., Palmesino, J., Rönnskog, A. S., . . . & Wolfe, A. P. (2017). Scale and diversity of the physical technosphere: A geological perspective. *Anthropocene Review, 4*, 9–22.

Zelazo, P. D. (2006). The Dimensional Change Card Sort (DCCS): A method of assessing executive function in children. *Nature Protocols, 1*, 297–301.

Zerubavel, E. (1989). *The seven day circle: The history and meaning of the week*. University of Chicago Press.

Zhu, Z., Dennell, R., Huang, W., Wu, Y., Qiu, S., Yang, S., . . . Ouyang, T. (2018). Hominin occupation of the Chinese Loess Plateau since about 2.1 million years ago. *Nature, 559*, 608–612.

Zietsch, B. P., de Candia, T. R., & Keller, M. C. (2015). Evolutionary behavioral genetics. *Current Opinion in Behavioral Sciences, 2*, 73–80.

Zimbardo, P. G., & Boyd, J. N. (1999). Putting time in perspective: A valid, reliable individual-differences metric. *Journal of Personality and Social Psychology, 77*, 1271–1288.

INDEX

Thomas Suddendorf is a professor in the School of Psychology at the University of Queensland, Australia. He is the author of the critically acclaimed book *The Gap: The Science of What Separates Us from Other Animals* (Basic Books, 2013). Suddendorf pioneered the study of "mental time travel." His work has been featured in leading scientific journals including *Science* and *Trends in Cognitive Sciences* and in popular outlets including *Scientific American* and *New Scientist*. He lives in Brisbane, Australia.

Jonathan Redshaw is a postdoctoral fellow at the University of Queensland. He has published extensively on the development and evolution of mental time travel, focusing on how children and animals think about uncertain future events. He was named a 2021 Rising Star by the Association for Psychological Science. He lives in Brisbane, Australia.

Adam Bulley is a postdoctoral fellow at the University of Sydney and at Harvard University, where he researches the cognitive science of foresight and decision-making. He has won numerous honors and awards for his research and teaching. He lives in Sydney, Australia.